A Telecom Library, Inc. Book
Published by Telecom Library, Inc.
 Published in the United States by Telecom Library, Inc., New York.

ISBN 0-936648043-0

Manufactured in the United States of America

First Edition, December 1993
Cover Designed by Mara Seinfeld
Printed at BookCrafters, Chelsea, MI.

Dedication:

To Kathie for her undying devotion and patience during the creation of this book.

Acknowledgments:

Many people were instrumental in designing, ratifying, and documenting the Signal Computing System Architecture. The body of work represented in several of the more technical chapters was the result of contributions by product managers, engineers, and technical publications staff at Dialogic. Although everyone at Dialogic and the charter endorsers of the architecture had a hand in its creation, I would like to extend a special thanks to the following individuals:

Technical Writers
Andy Henderson
Mark James
Patrick Boyd
Ray Cyphers

SCbus chip (SC 2000) designers
Jay Gerheby
Don Messa

SC Message Bus firmware
Wai Chan

SCSA downloaders
Darran Hunt

ServerAPI
Paul Ranford

SCbus, SCxbus:
Pin Lo Chen

SCSA routing firmware
Mike McCafferty

SC Software (Drivers, DPI, API, ServerAPI)
Joe Grecco
Paul Bennett

SCSA Firmware
Dennis McRitchie

Overview & User Group
Bob Heymann
Charlie Walden
Jill Groff

A very special thank you to Harry Newton for contributing to and publishing this book.

TABLE OF CONTENTS

Foreword

by
Harry Newton

Once in a great while you're able to witness a great happening, or at least feel like you're part of something big. I've been a proponent and all-round evangelist for the idea of Computer Telephony going on 15 years now. There's been lots of talk during that time, but not since this past year have I seen such a mad scramble of computer companies, telephone switch manufacturers, software giants, and LAN gurus to embrace this thing called SCSA - the Signal Computing System Architecture.

This book is aimed at the computer and telephone professional who wants to become familiar with the nuts and bolts of how computers and telephones are merging. Some of the chapters are "primer" in nature, and still others go pretty deep into code examples and advanced Client-Server Telecomputing. Reading this book cover-to-cover may cause brain damage, so I wouldn't try it in one sitting. If you're a computer professional, you'll find the approach to telephones, networks, and switching systems to be perhaps the most refreshing look at this technology I've ever seen assembled.

If you're a veteran to voice processing, this book takes you on a comprehensive journey through everything from host interactive voice response to advanced fax-on-demand implementations.

There's information on SCSA specifications as well as topology examples and profiles on the companies who are designing around the architecture.

With close to two hundred endorsers, top players in working groups, and dozens of associated projects, SCSA is quite simply put an all-out revolution to make phones, PCs, LANs, and everything else finally communicate. SCSA may very well fulfill the promise of surrounding customers with instant information. Now it's possible to coordinate order entry screen transfers with telephone calls, manage mixed media on the same platform, and unify messaging elements from the desktop.

I predict that within several years of this writing, SCSA and its endorsers will have caused an absolute avalanche of high-productivity Computer-Telephony products and services. Of course, you'll be able to pick-up the most substantive and newsworthy items by subscribing to Computer Telephony Magazine, which we kicked-off at NetWorld last month. Judging from the enthusiastic response from the readership of that magazine, we've really hit the mark in targeting the merge of these two worlds of computers and telephones. There's Billions of dollars just waiting to be scooped-up by developers, integrators, manufacturers, and programmers who adopt the precepts of this new architecture.

This is a technology that I am proud to have had some influence on throughout the years. It's absolutely gratifying to see the advances in services and products that are promised by SCSA. We're in for an exciting ride. I hope you find it enjoyable and a comfort to have this definitive guide to help you along the way.

Harry Newton
New York City, Fall, 1993

CHAPTER 1 - An Overview Of Voice Processing

What Voice Processing Does for Modern Communications.

Why should you as a communications or computer professional take time to understand what voice processing is and how new standards are affecting it? Because voice processing is revolutionizing the way we communicate, and if you understand it you can:

- Save thousands, even millions of dollars in people and equipment related costs
- Design enterprise-wide systems for better communication
- Turn your corporate database "inside-out" so customers can benefit
- Simplify access to previously difficult to reach information
- Win new customers and keep existing ones with a high grade of service

In order to better understand how you can benefit from Voice Processing and new standards arising from the voice processing industry, it is important to have a basic understanding of telephony and computation. This chapter will explain how the telephone system can be made to interact with computers with voice processing.

What's Voice Processing

Voice Processing is a term which is applied to a set of technologies that allow people to use regular telephones to access computer databases without a modem hook-up or terminal. With Voice Processing, the telephone *is* the terminal. In a sense, the keypad of the telephone is the "keyboard," and the receiver is the "screen." You could compare the discipline of voice processing to a protocol converter, because it allows two dissimilar devices to communicate.

For example, in the world of data communications, a protocol converter might be used to convert input from a regular ASCII terminal into output suitable for an IBM mainframe. A number of devices like this are available, either in the form of "black boxes," or in the form of PC expansion cards. A protocol converter, in this case would emulate the functions of a cluster controller on the "computer" side, and a data mux on the "terminal side."

FIGURE PROTCONV.DRW

PROTOCOL CONVERTER EXAMPLES

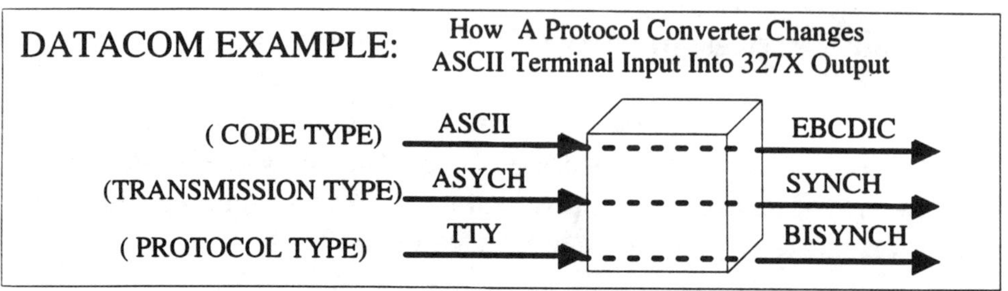

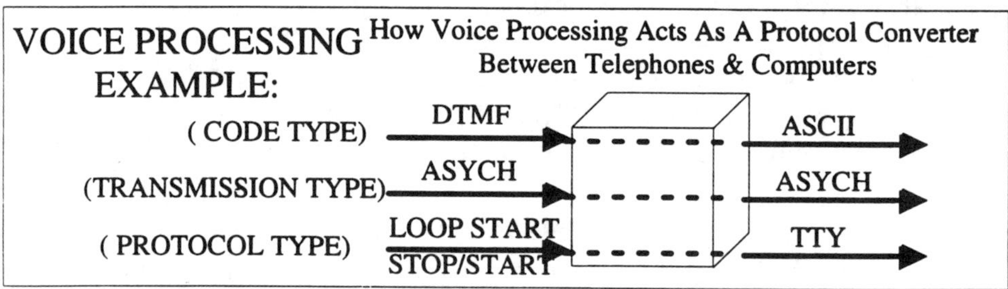

Voice Processing follows the same logic for protocol conversion as in the world of data communications. Figure (PROTCONV.DRW) shows how a voice processing device accepts DTMF (Dual Tone Multi-Frequency) code and then converts it into ASCII code, for example. This ASCII code could then be sent "upstream" to a transaction processor, or server, or other computing device for further action. In addition, the Voice Processing device can convert the human voice into a series of "ones and zeroes" for storage and later retrieval by using digitization techniques as explained later in this chapter.

In order to use the touch-tone pad as a "keyboard" for voice processing equipment, the voice processing gear has to "listen" for the tones and then translate them into computer commands. A variety of both discrete DTMF (Dual Tone Multifrequency, or Touch - Tone), and DSP-based (Digital Signal Processing) technologies are used to listen to the tones, much in the same way as the telephone company listens to the very same tones in order to route your telephone calls.

Your telephone pad represents a gird of intersecting high and low frequencies. The combination of these frequencies at the same time provides a combined (dual) tone. This dual tone is more difficult to "fault" than just one tone, so it is pretty reliable for directing phone traffic. Figure (CH1FIG6.DRW) shows the way your telephone keypad is "mapped" to generate these discrete dual tones.

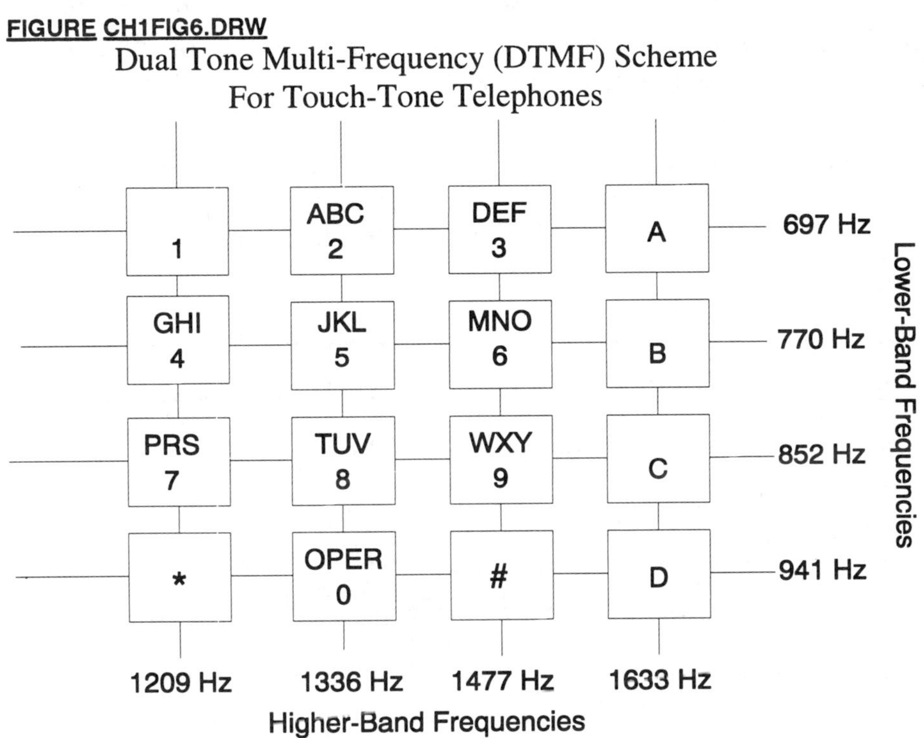

The voice processing equipment is programmed to accept certain series of these same tones at your command. Usually, you are "prompted" by a recorded voice to type in the DTMF tones as a way of responding to questions or instructions. In some cases, you are asked to leave a message for later retrieval, like you would do with an answering machine.

In most cases, the voice you are hearing from a voice processing system is a regular human voice that has been recorded, and then digitized for storage on a computer's hard disk. This voice recording, called digitization, is the preferred method for voice processing today of storing and forwarding the human voice in virtually every application presented in this book. There are a variety of encoding schemes to digitize the voice, however, they all basically do the same thing: apply a digital value over time to the varying amplitude of your voice.

As you speak, the amplitude of your voice goes up and down. An encoding scheme, or algorithm is used to assign numbers to the points of amplitude as your voice is recorded.

Figure (CH1FIG0.DRW) shows how this is accomplished with PCM, or Pulse Code Modulation, which is a popular method of digitizing the human voice over telephone circuitry.

FIGURE CH1FIG0.DRW

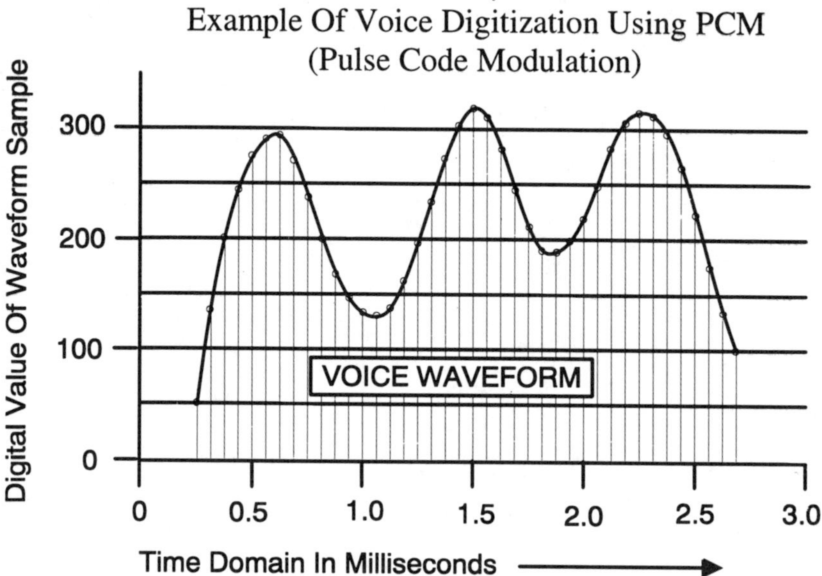

By combining the core technologies of voice digitization, DTMF encoding, and telephone line signaling, voice processing systems can be made to perform some pretty complex tasks that we are beginning to take for granted. For example, this technology is at work providing "bank-by-phone" services, which allow you to access your bank balance over the phone. The technology can also be found in voice messaging systems, which allow callers to type-in extension numbers to direct telephone calls and leave messages without the aid of an operator. These are just a few examples of how voice processing technology help us to "self-navigate" through computer-based information.

We encounter voice processing technology with virtually every calling-card type long distance call that we make. For example, when you make a long-distance call from a pay phone and are prompted to enter your account number with the "bong" tone, you are being guided by voice processing. Any pre-recorded voice which asks you to hold on before a "live" agent is available is an example of voice processing. There's simply no escaping it; voice processing is a reality of modern-day telephone calls.

Figure (CH1FIG1.DRW) provides an overview of how these technologies come together over the telephone network. In this diagram, a regular telephone can be used to call a computer which is hooked-up to a "Voice Processing Unit." The Voice Processing Unit converts touch-tones and other commands from the phone such as the spoken word into commands that are sent upstream to the Host Computer. These commands are processed as if they were received by another computer. The host computer responds to the Voice Processing Unit, which in turn "speaks out" information to the caller over the phone.

Even in its simplest form, Voice Processing has been around for almost as long as the modern telephone network itself. The Audichron Company was one of the first to fashion a mechanical device to the telephone network to play-out simple sponsored announcements for the time or weather. These systems have become much more sophisticated and now handle special intercept treatments on the "wrong number dialed."

The first time and weather machines were devices which employed a series of mechanical relays which literally lifted regular telephones "off-hook" when the sound of a phone's ringer sensed a caller on the line. This set off a series of actions that a 78 rpm record could be played over the phone with pre-recorded messages on it.

Hundreds of newer devices that use synthetic voice coils are still in use, but the most used method is solid state or computer-based voice digitization, as explained above.

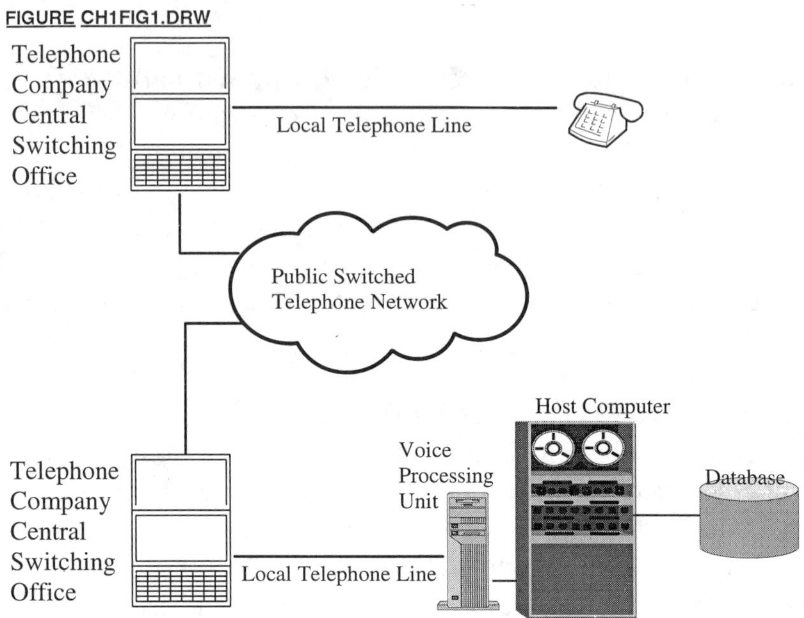

The Telephone Network

Since Voice Processing brings together the telephone world and the computer world, it makes sense to discuss the telephone network in order to explain the technology elements represented in voice processing.

What a mysterious animal the telephone network is to most folks... It's this amorphous cloud that everything form data traffic to voice communications are thrown into once a communication leaves your facility. It's not really as magic as most people make it out to be: In most cases, you telephone or modem is hooked-up to a pair of wires that disappear in the wall or snake around to a pole outside of your home or office.

These wires are but a few in literally thousands of bundled pairs of wires that are servicing your community. The single pair of telephone wires connected to your phone or modem can be traced back to a "line finder" in your local telephone company central office. A central office is usually housed in a brick building with a telephone company insignia on it (usually a picture of a bell of some kind), with small shrubs and parking enough for one medium-sized utility vehicle.

A central office contains a switching matrix, a cross-connect frame for physical wiring, and a wall of batteries to power the telephone lines. Telephone calls require "talk battery", or 48 volts of electricity to carry your voice over the long distances required to get your telephone call form your home to the central office.

These batteries are constantly "trickle-charged," so that the telephone traffic will flow even in the case of a power outage. This is why your telephone always seems to work even when the power is out during a storm.

Figure (CH1FIG2.DRW) describes the "POTS," or plain old telephone service that is provided to most homes. Your telephone (customer premise equipment) is attached to a "local loop" pair of wires, that provides enough energy to carry a conversation and to transmit tones and other information to the telephone company central office.

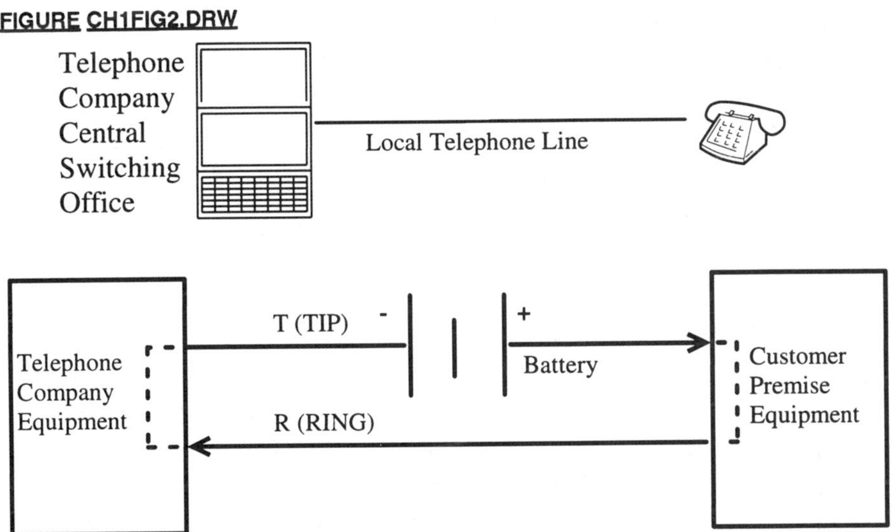

Diagram of a two-wire interface utilizing loop signalling
(POTS or Plain Old Telephone Service Line)

Other communications gear is connected to the telephone network in a similar fashion. For example, PBXs (Private Branch Exchanges) or Key Systems - the business telephone systems that we use in offices - uses

POTS lines as well as a variety of more sophisticated telephone circuits. These other circuits include DID (Direct Inward Dial), E&M TIE Trunks, and Digital Communications Lines like T-1 service.

A PBX is used to allow many persons at a company site to "share" a group of telephone lines coming into a building. That's why PBXs are referred to as "switches." Telephone lines coming into a PBX from the telephone company central office are sometimes called "trunks," and the telephone lines connecting the PBX to the phone on a person's desk are usually called "lines." Figure (CH1FIG3.DRW) diagrams a typical PBX configuration: A number of trunks connect the switch to the central office, and the telephone extensions connected to the PBX have access to those trunks via the station line connections.

FIGURE CH1FIG3.DRW

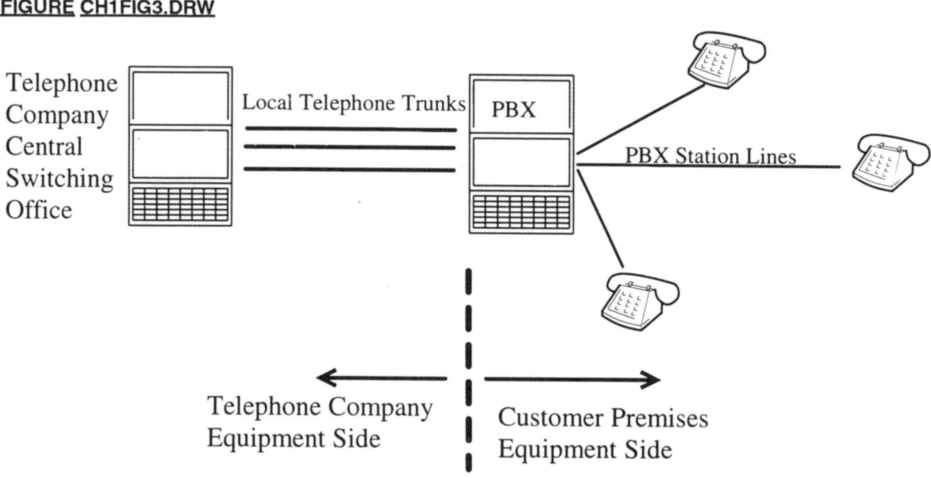

Larger companies sometimes "tie" their PBXs and other telephone systems together in a network. The trunks between PBXs in a network are often called "TIE lines" or TIE Trunks. The purpose of a tie line is to allow users on each PBX to call one another by simply dialing a prefix in front of the called party's extension number. Dedicated communication lines between PBXs are sometimes economical, because the cost to lease a full-time line or lines between the PBXs can be less expensive than making a long distance call each time one you make a call "outside" of the PBX you are calling from.

TIE trunks require more sophisticated signaling methods in order to ensure that calls don't "collide" with one another. Regular telephone lines have difficulty in getting around this problem, as we all have experienced "going off-hook" to make a telephone call, only to find that someone is already on the line calling you... To get around this problem, telephone companies use a number of methods. A popular method for analog telephone lines is called "E&M" signaling. E&M signaling uses the same pair of wires to carry the voice communication as do POTS lines, however, an additional pair of wires is used to supervise the call. This call supervision is sometimes called call "setup" and "tear-down."

Figure (CH1FIG4.DRW) details the way in which two geographically separated PBXs can be "tied" together with TIE trunks. A station behind one PBX can dial a digit which is read by the switch as a command to make a telephone call to the other switch. In telephone lingo, this is called "seizing a line." The first switch is called the originating switch in the case of the first person going off hook to make a call to the other PBX. Upon receiving the command for the local station, the originating switch applies battery current on the "M" lead of one of its Tie Trunks.

FIGURE CH1FIG4.DRW

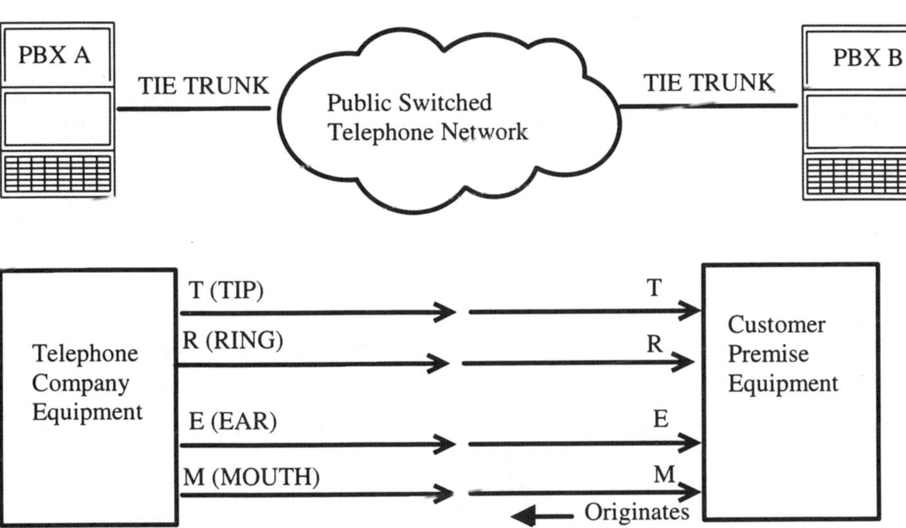

Diagram of a two-wire interface utilizing Type 1 E&M signalling
In this example, the Customer Premise Equipment originates a telephone call by applying "battery" to the "M" lead. This is a typical circuit used to connect PBXs together with "TIE TRUNKS."

This battery seizure is sent through the telephone company central office to the remote PBX. The remote PBX senses the seizure as a request for service, and then it "winks" back by toggling the Tip and Ring circuit. In effect, the two switches are setting-up an "ack/nak" handshake in order to establish a telephone call. This handshake differs depending on the type of trunks circuits that are engineered and the type of equipment used on either end.

After the handshake is established, the originating PBX forwards the digits that were dialed (usually and extension number) by either pulsing or sending DTMF over the (T&R) circuit. At that point, the telephone call is routed by the receiving PBX to the correct extension number.

This E & M type signaling has been so widely used, that the same basic call set-up and tear-down rules are used for digital telephone services such as T-1 and ISDN (Integrated Services Digital Network). This is significant for this book, because Voice Processing Systems are used to emulate these type of circuits either as sending equipment, or receiving equipment. Voice Processing equipment that does this emulation is also used by telephone companies to aid in the trafficking of phone calls, both by analog (E&M) or digital (T-1) means.

Digital T-1 service takes the equivalent of 24 analog (POTS) telephone communications and bundles them together over the same pair of wires using TDM (Time Division Multiplexing). The speed of a T-1 circuit is 1,544,000 bits per second, or roughly 64 kbps times 24. 64 kbps is the bandwidth required to carry the human voice.

This same circuit can bee fractionalized to carry lower speed communications such as data modems and fax machine traffic as well. Regular T-1 circuits utilized unused bits of information in the communication channel as signaling information instead of using dedicated pairs of wires as in the analog E&M trunks. These signaling bits are "robbed," out of the stream of available data in order to achieve the call set-up and tear down required to supervise telephone calls.

A signaling bit is either "high" or "low" as a 0 or a 1. The presence of a 0 or 1 during the setup of a call indicates whether or not the line in question is "ringing," "going off-hook," "hanging up," and so on. Since regular T-1 lines uses these bits for signaling, they are not available to carry information. In the case of a regular voice communication, that's quite all right, because we humans can't really tell the difference if a few bits are lost here or there. Data communication equipment, on the other hand is quite a different matter. This explains why SNA networks and other devices that use 56 KBPS data lines cannot utilize the entire 64 Kbps out of a T-1 span: Because the effective data rate after accounting for signaling bits is 56 vs. 64 Kbps.

ISDN circuits use the same basic digital facilities, however, all of the signaling is dedicated on a single data channel instead of "robbing" signaling bits from each communications path. That's what is implied when you hear the term 23B +D. The translation of that term is 23 bearer channels (lines that information is carried on) and one data channel (dedicated to handle the signaling and supervision of all of the bearer channels).

Relevance to Telephony & Computation

Voice processing is not the only discipline to combine telephony and computation. Figure (CH1FIG5.DRW) shows how T-1 multiplexors can be used to aggregate traffic from separate devices in order economize on leased telephone and data circuits. In this example, an enterprise may have two geographically separate sites with both telephones and computers which must be connected at either end.

Some companies still lease separate data lines to connect their terminals and pay for separate lease circuits for their regular telephone traffic. By adding a T-1 mulitplexor, or other routing device, both data and voice circuits can be combined and sent over the same telecommunications facilities.

In this example, "PBX A" does not have any data traffic running through it, and it is hooked-up to the T-1 multiplexor in parallel with the Front End Processor. "PBX B" has both data terminals and telephones hooked-up to the station side of the switch, so the PBX itself is acting as a T-1 multiplexor.

FIGURE CH1FIG5.DRW

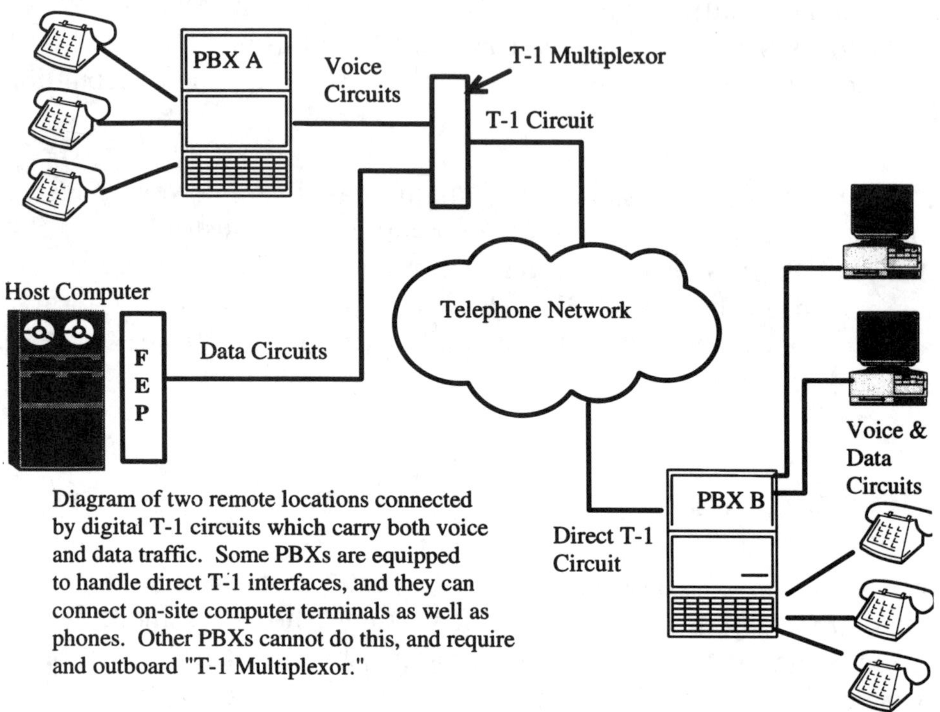

Diagram of two remote locations connected by digital T-1 circuits which carry both voice and data traffic. Some PBXs are equipped to handle direct T-1 interfaces, and they can connect on-site computer terminals as well as phones. Other PBXs cannot do this, and require and outboard "T-1 Multiplexor."

The relevance of T-1, ISDN, protocol converters, 56K links and other data communications elements is the fact that voice processing technology borrows heavily form these technologies. In fact, voice processing is a way in which these disparate technologies can be used together in powerful ways.

As an industry in its own right, voice and call processing is producing over $2 Billion a year in factory shipped revenues. About ten years ago, call and voice processing was just a fringe technology, but now it touches our lives every day in the form of voice mail, automated attendants, home banking, and pay-per-view services. Some say that call processing is the business of "making things work together that weren't made to work together."

The Mix Of Computing Models, Operating Systems, and Platforms

Companies involved with voice and call processing are faced with some unique challenges because of the fact that they straddle the fence between computing and telephony. Call processing systems are sometimes treated like a hot potato, in which only the ultimate success or failure of a new system determines the ownership between the computer folks or the telephone folks. Since call processing is a subset of telecommunications *and* computation, players have to possess competence in a variety of disciplines including LANS, WANS, Digital Signal Processing, Switching, and Microprocessors. The technology landscape is changing so quickly, that it's very difficult for us to keep up with the rapid pace of change in technology.

Software, too, plays a vital role in the way in which all of these disciplines weave together. For example, the outcome and acceptance of Microsoft's proposed T/API (telephony application programming interface) may well be the bellwether for a whole new age of communications. An age of cooperative computing and interchange between separate systems. Although the T/API specification defines little more than making phones go on hook and off-hook, it is nonetheless a peek at how these two worlds are coming together. Even UNIX, the bastion of telephone companies, is getting into mainstream computing via Novell.

Another layer of complexity in our marketplace is the disparate form factors of computers. We have ISA, EISA, NuBus, VME, MultiBus, FutureBus, and a host of proposed buses that are pretty hard to keep up with.

The mix of computing models...

In terms of topologies and computing models, there are a variety of configurations that have been used in Call Processing:

- Self-Hosted
- Host-Slave
- Client-Server

For example, there's the "classic" Self-Hosted mode of voice computing, where the application, line interface, signal computing, and other related hardware all reside inside of the same box. This is manifest in the majority of CPE-based Voice Mail and Audiotex systems, for example.

Host-Slave models use voice processing systems as peripheral devices, and the application software resides in a larger host computer - either a minicomputer or a mainframe. In this model, the voice/call processing system is, in a sense, a remote I/O device, which acts as a "*protocol converter*" between the world of telephony and the world of computation.

The client-server model uses a mixture of the first two ideas. For example, a client computer, such as a personal workstation may have some software in it that allows the user to "call-up" the use of a centralized video, fax, or voice services capability which resides in a "server." The server machine can have an array of technologies in it that are used as a centralized resource to all of its clients. This is typical of workstations sharing a printer through a "print server" over a LAN.

The Predictability Of Commoditization

High-tech companies have found keeping up with all of the advances while still being sensitive to their installed base is very difficult. At issue with all of the players in communications and computation is the concept of predictability. So, over the years companies have done a number of things as a hedge against all of the changes, and that's to form strategic relationships and agree on standards. This helps us all to better anticipate market demands more effectively than alone. This is a key concept, because no matter how brilliant a company's research and development staff is, they quickly discover their limitations in terms of the ability to always be ahead of the "power curve" with the latest and greatest technology.

Although each products company needs to have a severe commitment to its core technology, customers always need complimentary technologies such as Text-To-Speech, Switching, Voice Recognition, Video, and others in order to be successful in their markets. It isn't always obvious to a company that in order constantly create new and better things for customers, that you run out of people and resources in short order.

What seems obvious on the surface is that computer and communications companies should have been sharing technology and research with one another over the years. That the best course of action was to share technology with other developers and manufacturers. The computer industry has been doing this, albeit reluctantly, for about ten years. The telephone industry is lagging behind for the moment.

When computer and communications companies get together to solve customer problems, they discover that customers want more than computers and a telephone switch. Customers want to make it all into one system and also have access to Text-To-Speech, Voice Recognition, Fax, Video, and Voice Processing. These technologies use a mixture of discrete and DSP-based components, although it is clear that DSP is assuming a leadership role in the advance of core technologies.

The advances in core technologies and DSP...

Take facsimile as a case in point. We are witnessing the speeds of a fax transmission getting faster and faster. Now there's a new speed that everyone wants to use which is 14.4 kilobits per second, which doubles the transmission speeds used just recently. And before most developers come up to speed technology will have advanced even further. This is fueled by users' thirst for "*Higher Speeds And Better Feeds.*" *Feeds* refer to the interoperability between machines and the ease with which disparate technology elements talk to each other.... and *Speeds* are how fast things happen. But these speeds and feeds keep changing at a dizzying pace.

Core technology vendors sell DSP algorithms that perform fax, video, voice digitization, and other task. In the past, these vendors would sell their signaling algorithms to OEM customers, who would bundle the technology and sell it on proprietary platforms.

The business model for core technology access is beginning to change because of Open Systems Development. More and more, algorithms are becoming available in development packages, so that a higher number of open systems developers can take advantage of the technology and get it to market quicker.

Intel, VLSI, Motorola, and Analog Devices are a few of the chip manufacturers that are supporting this new way of approaching modern signal processing. Figure (ADI01.DRW) displays how "Technology Vendors" are beginning to broaden their access to market by developing their technology on standard platforms.

FIGURE ADI01.DRW

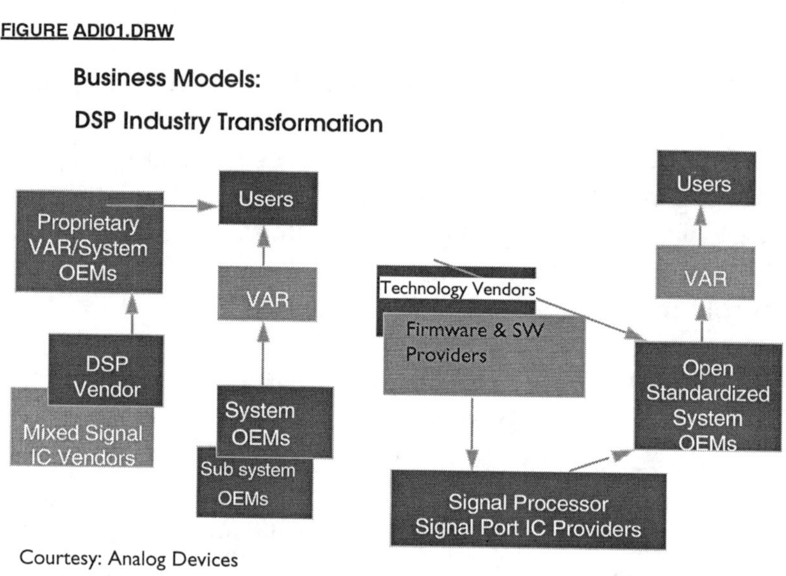

Courtesy: Analog Devices

Some of the leaders in the industry who have mastered these DSP technologies include Dialogic, Voice Processing Corp, Scott Instruments, GammaLink, Biscom, BrookTrout, and Berkeley Speech Technologies to name just a few. Some of these products are cross-licensed back and forth, either by reselling components or licensing algorithms that work on different platforms.

Now what's happening is that the voice and call processing industry is entering into an age of absolute commoditization. And the whole landscape has changed to the extent that other technologies that cross over into the call processing area have reached commoditization, too. This includes modems, fax, voice recognition and video. Now you can get a fax board for less than $100. All because of the prolific PC. After additional technologies get the bugs worked-out, they all end up on the PC form-factor sooner or later.

And now, the idea of mixing real-time video is becoming a marketplace demand. The problem is that when you really look at multi-port implementations of these other technologies, and the bandwidth they require on resource sharing busses such as the PEB (PCM Expansion Bus) or the MITEL ST Bus, we simply run out of gas.

People want to be able to take captured video data transmissions and send them inside the PC across a mezzanine bus, so that you can consolidate video messages and voice messages and regular telephone calls (and consolidate all that traffic down onto the same telephone line). This requirement stems from the desire to put all of the technologies into one platform, rather than having to deal with the complexity of disparate machines which may or may not be truly interoperable.

Commoditization is inevitable in any successful industry. It doesn't matter whether it's computer aided design (CAD), or if it's word processing. Look at what happened with word processing and leaders in that field such as Wang, for example.

You'd start out with something proprietary; something that the only works according to a single technologist's vision. In the case of word processing, a large box sat there attached to large terminals with oversized cables that connected to each terminal. It was this magic thing called word processing which was to have signaled the death knell of the common typewriter. It was going to eliminate typewriters, right? And thousands of companies invested in them.

But then the PC was introduced in 1979 and a revolution began. A few years later, it seemed that every major company was buying PCs, and there were a couple hundred word processing packages that became available either commercially or through "freeware." They were kind of clunky. But eventually, thousands more people bought these PCs and then eventually hooked them up to local area networks. People could easily share files back and forth with one another. It wasn't long before these big, bulky, proprietary systems went out of style.

History is repeating itself in Call Processing...

History is repeating itself with voice and call processing. Voice processing really hit the "big time" in 1979 with the introduction of VMX's Voice

Message Exchange. It had a proprietary operating system, proprietary hardware, and a non-portable application software package. But as time went on, other companies produced systems that de-coupled the application from the voice processing platform, like Voicetek, for example.

Then, an era of renaissance occurred in 1983, when Dialogic introduced a PC form-factor voice processing card. This enabled developers to build sophisticated voice processing systems where the hardware, firmware, platform, and application software could be provided by individual contributors.

This separation of expertise in technology vendors is called "Value-Added Granularity." Figure (BEFORE1.DRW) shows from left to right, how open systems design was preceded by more proprietary systems, that required "hard coding" in order to deliver applications to the end user. Some older voice processing systems were designed as proprietary because there was no choice (because off-the-shelf technology was not available), and despite a wide array of economical off-the-shelf components on the market, some companies still insist on creating one-off designs for proprietary systems.

FIGURE BEFORE1.DRW

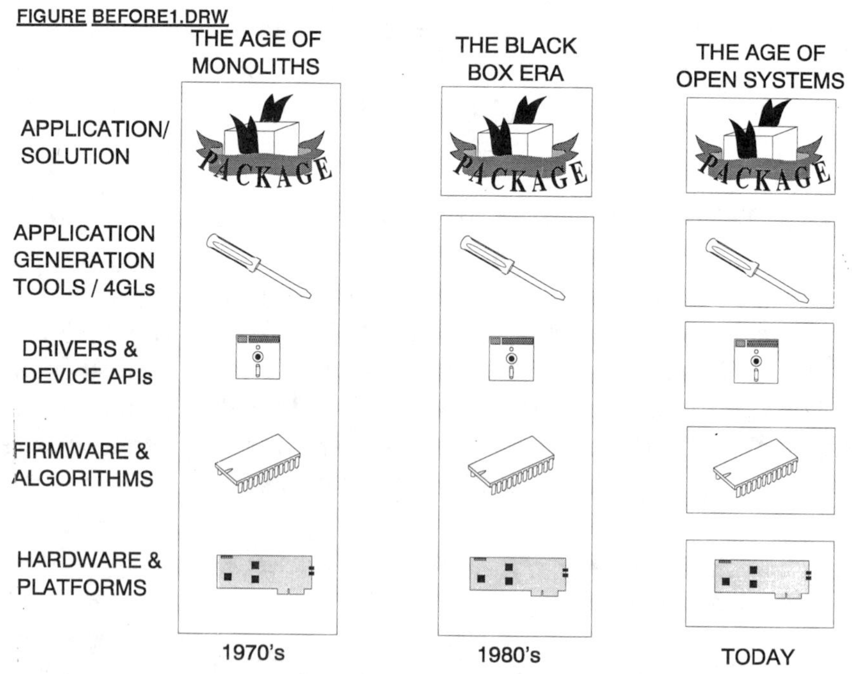

The trend of open architecture is changing all of that. Some of the biggest players in Voice and Call processing today, including IBM, Tandem, Brite Voice Systems, Perception Technologies, and VA all use open-architected systems. In fact, both Octel and VMX, the two largest proprietary hold-outs in the industry have recently made technology and company acquisitions of an open-architecture-based nature.

Octel, VMX, AT&T, ROLM, and Northern Telecom are the clear industry leaders from a revenue standpoint, and Dialogic is the marketplace leader in voice and call processing from a "ports shipped" and enablement standpoint. The company's technology has helped to launch literally hundreds of companies into the industry. The following table highlights some of the milestones over the past ten years in PC-based voice and call processing:

1985 Significant Milestones

•	First DOS-Compliant TSR for Multi-port Call Processing Delivered
•	First 4-port Voice Processing PC Board (DIALOG/40)

1986 Significant Milestones

•	First Shared Resource Bus for PC-Based Call Processing Development (Analog Expansion Bus)
•	First Analog Matrix Switch for Integrated Voice Processing in PC (AMX/xx Audio Multiplexor)

1987 Significant Milestones

•	First Viable Voice Recognition Product for over-the-phone use on PC (VR/10)

1988 Significant Milestones

•	First Backwards Compatible OS/2 Upgrade for Installed DOS boards (OS/2 Voice Development Package)
•	First multi-port daughtercard for call processing to enable ANI in PCs (MF/40)
•	First UNIX compatible driver for AEB-based voice applications (UNIX Voice Development Package)
•	First Digital Trunk Interface for the PC eliminates need for channel banks(DTI/1xx)

• First to achieve 16 voice boards per PC now viable using patented shared interrupt scheme

1989 Significant Milestones

• First PC-based Platform to meet LSSGR/NEBS compliance for telco use (DTP)
• First 12-Line Voice Processing Board for the PC (D/12x Voice Communication Platform - Spring Board)

1990 Significant Milestones

• First PBX integrations compatible with industry-standard Voice APIs (NTI SL-1 Integration & Mitel SX/xxx Integration Cards)
• First High-Density Fax Solution in the Industry for the PC (FAX/120 12-Channel PC Fax Board)
• First Open-Architecture 8-channel VS&F capability for IBM's PS/2s (D/81MC, LSI/80-MC, VR/41MC)
• First Call Processing Industry Co-Marketing Program of 4GLs for PC (ToolKit Developer Program)

1991 Significant Milestones

• IBM's Call Path Services Environment goes "PC" (IBM DirectTalk/2 using Dialogic Cards)
• Industry's first PEB-Based Digital telephony Interface for Europe (DTI/212)
• Over 100 licensees join to work on AEB & PEB design packages & tools (OPEN Development Program)
• First unified release of 4 and 12-port Recognition for AEB & PEB products (VR/40 & VR/121)

1992 Significant Milestones

• First multiport voice processing environment for AIX for use with industry standard voice cards (AIX Voice Development Package)
• Industry's first 4-port fax daughtercard compatible with installed vs&f software (FAX/40 daughtercard)
• Vicorp BETEX-ESP first commercial application of Dialogic's AppServer On Tandem (Tandem announces Tandem Non-Stop Call Center)

• Dialogic Introduces AppServer API for Remote VRU control (Host/Slave Model)
• Dialogic/Northern Jointly Announce Development of Integration Products For Norstar (D/42-NS)
• First generic baseline development platform for PBX integrations (D/42 Integration Platform)
• First enhanced DSP package to run on multiple low & high density platforms (Introduction of Springware - Scalable firmware for DSP-based Voice Processing)
• First PEB-based ISDN interface for international use (PRI/211)

FIGURE BUBBLE.DRW

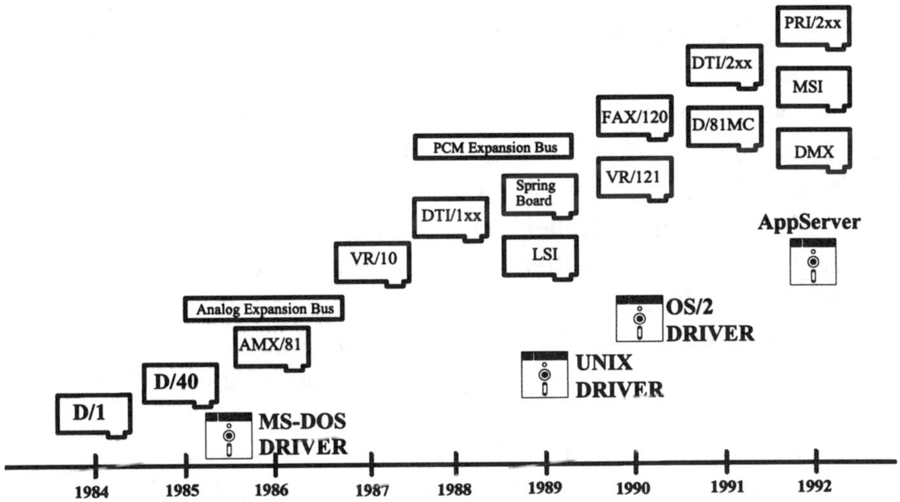

1993 Significant Milestones

Dialogic Announces Signal Computing System Architecture - SCSA

... Over seventy industry endorsements from Computer, Telephony, and Call Processing sectors ... most comprehensive standards-based multi-media development initiative includes the third generation PCM/TDM BUS, out-of-box transport, unified drivers and firmware specifications, and Application Programming Interfaces for a variety of computing models and configurations.

As you can see, the use of PCs in voice processing has had significant impact on the growth of the voice and call processing industry. Initiatives such as The ***OPEN Development Program***, which was spearheaded by

Dialogic, and MVIP, spearheaded by Natural Microsystems, sought to bring together Solution Developers, Technology Developers, and makers of Application Generators in 1990.

Hundreds of developers use these standard APIs, driver conventions, and resource-sharing busses, including AEB (Analog Expansion Bus), PEB (PCM Expansion Bus) and MVIP (Multi-Vendor Integration Protocol, based on Mitel's ST-Bus). Even Dialogic's chief competitors Rhetorex and Natural Microsystems are using the AEB. In fact the drivers and APIs based on the popular D40LIB (the "C" libraries that come standard with all Dialogic voice cards) are the most emulated by Dialogic "clone" competitiors.

Existing architectures, both Proprietary and Open

How do we get all of this complex technology to work together? Millions of users are in need of these disparate technologies, but they view it as a big mess. From a historical perspective, a second-generation architecture for binding call processing elements together was defined by Mitel and Natural Microsystems. It was called MVIP which stands for Multi Vendor Integration Protocol, and it was a step in the right direction. MVIP uses a second-generation bus akin to Dialogic's PEB (PCM Expansion Bus) that was promoted as a way to link all of these technologies together. It's important to note that the penetration of PEB-based systems using this same concept were already being deployed by the thousands. The basic idea was: "Hey, let's develop some kind of a mezzanine bus that connects PC cards." This mezzanine bus is a bus that goes on top of the cards in a PC.

So, both the PEB and MVIP busses are used for transferring information back and forth between resource cards, so that the cards can share information and maybe switch a telephone call back and forth on that bus.

The concept was to develop a standard that various manufacturers of core technologies could conform to: "Here, you get your bus to look like ours and we'll get ours to look like yours and we'll share the specifications. This way we can get our products to market quicker together." So MVIP supporters mimicked the PEB concept and touted it as an open architecture for the voice processing industry. The primary difference, however, was the way in which MVIP constituents promoted the bus *as an open standard*. The Dialogic PEB specifications were made public just months after the MVIP announcement, and so began a revolution that eventually brought us the breakthrough of SCSA (Signal Computing System Architecture), which is what this book is all about.

SCSA seeks to erase the ambiguities in interoperability and to demystify the complexity in bringing the computer and the telephone together. PEB is the first generation PCM/TDM mezzanine bus for voice and call processing. MVIP is a second-generation bus which has a higher bandwidth than PEB. Within the SCSA architecture, the SCbus is vastly superior as a third generation bus, although in certain modes, it is backward compatible to the first and second generation busses.

The industry needs SCSA...

The industry is in dire need of rationalizing these earlier generation busses so that we can move forward and take advantage of newer technologies that need higher speeds and feeds. On the other hand, SCSA supporters don't want to abandon loyal customers who have made huge investments in either PEB or MVIP-based systems. What could possibly be better than what they're doing? What really turns people on? Speeds, feeds, compatibility, and above all: *STANDARDS* which supply some kind of predictability in design.

Developers are demanding not only bus standards, but also standards that have to do with software application program interfaces as well. This is where the second generation architecture really falls short of vision. In addition, the second generation architecture lacks cohesive access points and definition on driver standards and firmware development kits.

CHAPTER 2 - Voice Processing Applications

There are a number of ways to look at applications and solutions. For example, you can take generalized functions that solve a set of generic problems, and take a "horizontal" view of applications. You can also take an industry-specific or "vertical" view. In addition, you can look at a topological, or environmental view. In this chapter, all three views will be touched on, starting with an overview of horizontal application segments:

Horizontal Application Segments In Voice and Call Processing

- Messaging
- Interactive Voice Response
- Integrated Voice & Fax
- Operator Services
- Telemarketing
- Audiotex

The actual environments that these applications are placed in boil down to either on-site or off-site implementations. In the case of voice processing gear, it's either CPE (Customer Premise Equipment), or telephone network-based gear. The table below typifies the type of technologies which are co-located in either environment.

Computing and Telephony Environments

ENVIRONMENT:	TYPIFIED BY:
CPE	Standalone Service Bureau Systems PBX, ACD, or Key System Adjuncts LAN-Based, Server Orientation
Network	Telephone Company Central Office Long Distance Carrier Class 4 Tandems Private Networks Cellular and Radio Paging Systems

Whether CPE or Network-based, the distribution of ports over the horizontal application segments have been fairly steady over the past three years. Messaging applications are clearly dominate, and heavily penetrated on the

CPE side. According to a number of sources, including Probe Research and Tern Systems, there were over $2 Billion generated this past year in terms of factory shipped revenue, with 30 to 50% growth forecasted over the next few years depending on the application segment.

Including both board-level and proprietary ports, the total number of lines shipped last year topped 800,000. PC-based systems accounted for close to 50% of all of these ports, making open-architecture a key engine of growth for the voice processing industry.

Percentage of Port Distribution Between Horizontal Application Segments

	1991	1992	1993
Messaging	46%	45%	45%
IVR	14%	15%	16%
Integrated Fax	1%	1%	2%
Oper. Svs.	7%	6%	4%
Telemktg	18%	18%	19%
Audiotex	13%	14%	12%

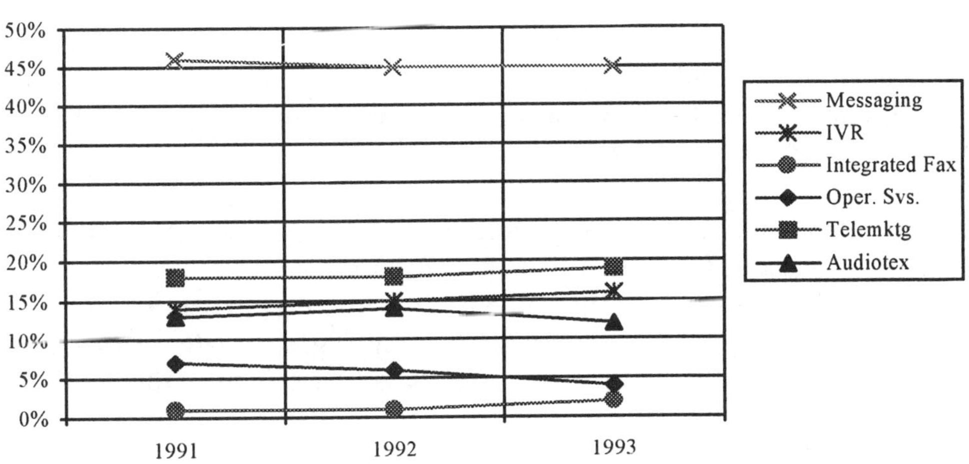

Messaging

This application segment includes all solutions for the store and forward or recording and retrieval of voice messages. These messages are recorded and forwarded in a non real-time way, much the same way an answering machine works. The difference is that messaging systems are a kind of "shared tenant" answering machine, because messages that were intended for as many as a thousand or more users can be stored and controlled by the same system. If a community of users agrees on some basic ground rules, messages can be shared, forwarded, and distributed to multiple recipients the same way electronic mail messages can.

The users can be individuals as well, so that messages received are dealt with in a simplified (answering machine-like) fashion. Users can be internal or external to an organization or both. When these messaging capabilities are centralized with a PBX, it is typical to have automated attendant functions as well.

An automated attendant will answer the phone on behalf of a live person as explained in the messaging example below.

Voice Mail applications are manifest in a variety of environments, including CPE (Customer Premise Equipment), C.O. (Central Office), and Network implementations. Examples of CPE-deployed Voice Mail include AVT's call express system. C.O.-based voice mail includes solutions such as Boston Technology's C.O. Access System.

Network implementations include nationwide paging with voice mail like Skytel, and a variety of Cellular Telephone based systems. In addition to voice messaging, this segment includes dictation applications as well. Dictation allows for the recording of spoken messages for later retrieval and transcription. Dictation solutions come in a variety of sizes, both for individual and centralized use.

Example of a typical messaging application: Corporate Communication

Corporate communication systems are busy hubs of communication. Because of peak calling traffic during the day, and also after-hours traffic, the reception and operator staff are often overloaded with calls. This can be frustrating for callers and employees alike. Callers are often put on hold several times and transferred to a number of extensions before they reach the intended party. This not only keeps the telephone lines busy, but also creates frustration for callers who may hang-up and do business elsewhere.

Call processing systems that solve these corporate communication problems are popular and varied. The most commonly installed is a combination automated attendant & voice mail system. When a caller dials a company with such a system, he or she is presented with options to either access the intended party through an automated personnel directory, or is prompted to type-in a digit which indicates what department is required. These systems allow the switchboard operator to concentrate on helping visitors or callers that require a human operator for call routing or directions.

In addition to touch-tone input, an automated attendant can be programmed to ask the caller to ***speak*** his or her name in order to prioritize, or screen the call so it can be handled better. Some more sophisticated systems will even prompt the caller for their customer number or even collect it automatically by using ANI (Automatic Number Identification). Customer numbers can then be forwarded to the company's computer system, so that a service screen can be transmitted to an agent's computer terminal in preparation of receiving the caller's transaction or request.

The modern voice mail system is the most common solution for corporate communications overload. These systems not only allow callers to leave detailed messages, but they can also attempt to locate and notify the recipient. This is achieved with the use of outdialing to beepers or other communications devices including paging systems. Voice mail is especially helpful after normal business hours, when no switchboard operator is available to alert employees about messages. Employees can leave detailed greetings which indicate when he or she will be back from a meeting or available. Additionally, these greetings can indicate travel schedules and other important information such as alternate phone numbers and fax machine numbers.

Traveling employees can pick-up and receive messages from any telephone, and they can respond to messages sent by customers or fellow employees by dialing digits or other commands indicating the treatment of the original message. For example, a message can be forwarded to another party with comments attached, or forwarded to a group of individuals for review. The same is becoming true for fax messages as well. In fact, some new systems allow fax messages and voice messages to be queued in the same "storage file" so users can retrieve information from any fax machine or regular telephone. Depending on the nature of the message, much of this can be done during the same transaction.

Interactive Voice Response

Interactive Voice Response (IVR) systems provide "transaction processing" or "transaction passing." Transaction processing systems not only act as front end, but also batch information that is later uploaded to a mainframe computer or other computer system in a standard format. Transaction passing systems use voice processing as the primary customer interface, but only "backwards engineer" or emulate what regular data entry terminals are doing for live operators. IVR examples include Pay-Per-View, Home Banking, Flight Status, Order Entry Systems, and, Inventory Control systems.

Example of a typical IVR application: Bank-By-Phone Service

Bank-By-Phone (home banking with voice processing) systems interface to mainframe computers that handle a customer's bank account transactions, and the IVR system is simply passing messages back and forth between the caller and the main frame. These systems handle customer requests for account balances, funds transfer, and loan information. Although ATMs (Automatic Teller Machines) are popular and provide access to cash, they don't typically provide the ability to carry on transfers, inquiries, and bill-paying services that are available through home banking.

Call processing technology can be applied to increase service levels for bank customers. Systems designed for home banking use terminal emulation cards or protocol converters which are commonly available. These emulation cards and converters act in the role of a bank teller's mainframe terminal, and are placed inside the same computer as the cards that make-up the call processing system. A bank customer will call a telephone number and then be prompted much the same way a bank teller would prompt the caller over the phone.

Security measures are still required for these transactions, so they are usually signed-up for ahead of time by the customer, so a PIN (Personal Identification Number) can be chosen by the customer.

These PIN numbers are used to validate the identity of the caller. In fact, so systems actually use "voice verification" technology, which matches key words spoken by the caller to a stored recording of the caller's voice. Companies like Voice Control Systems of Dallas, Texas are on the forefront of this technology.

Once the caller is verified, he or she is asked by the system to enter his or her account number and identification code, and then to choose from a variety of services. These services may include the quoting of the available funds in a certain account, or the outstanding balance of a loan. In some cases, a caller can ask the call processing system to respond with the "pay-off" amount of a loan. Many systems allow for the transfer money between from checking, savings, and other accounts.

Integrated Voice & Fax

Many voice processing systems are beginning to allow for the transmission of facsimile images. This is true of messaging, IVR, and server-based solutions. Facsimile-based solutions are defined as those applications where the chief purpose of the communication is to send or retrieve a scanned fax image to a caller or group of callers or recipients. An audiotex gateway that is primarily an interactive voice response system should not be classified as a "fax" system just because it has an ancillary fax capability. The service should be touted and billed as a "fax" system for it to be classified such. There will be some ambiguity here, inasmuch as MAP applications will cross-over into this segment and vice-versa. The 'ole "what's the dog and what's the tail" argument is, at least for the time being, a good way of reasoning this through on a case-by-case basis.

Each of the defined environments seem to be able to support fax applications. In the case of CPE, there are already stand-alone fax systems which broadcast faxes or support on-demand publishing of data sheets.

Fax applications will also be supported in C.O. environments for AIN (Advanced Intelligent Network) implementations. There are already a number of Network-level fax gateways, including both Sprint and MCI implementations. LAN-based environments support "server"-type fax stems, which allow a community of local users to share a centralized fax capability. Systems provided by companies such as FaxBack, CTI, and Nuntius are examples.

Example of a typical Integrated Voice and Fax application: Talking and Fax On Demand Yellow Pages

The Yellow Pages have long been the most reliable and comprehensive source for information about products and services that are available to consumers. The problem for users of these books is that the publication is only updated annually, so the information that is available is fairly generic and not very time sensitive. Voice and Fax processing have changed all of this, because now many "yellow pages" and even classified ads in newspapers have voice and fax ancillary functions.

These systems are called "Talking and Fax On Demand Yellow Pages." The only difference in the print ad is the appearance of a code number which appears somewhere within the advertisement. Consumers can call a general number and then type-in the code number corresponding to the product or service in question. These voice and fax provisioning systems are often front-ended by a custom message which can be recorded by the business who is advertising the product or service.

For example, the main menu might say: "Thank you for calling the Fax Pages. Please enter the fax code for the product you are interested in."

Once the voice processing hardware interprets the digits, the application directs the stored information to be sent over a fax card to the calling party's fax number. This can be accomplished in some cases by sending the fax during the same transaction if the calling party initiated the transaction from his or her fax machine.

This is where the term "one-call" and "two-call" comes from. A "one-call" design allows the fax transmission to occur during the same phone call. A "two-call" design requires the caller to either input the number of their fax machine, or otherwise identify him or herself so the program can send the fax to a predetermined number during a second phone call.

The next thing a caller may hear would be: "Thank you for asking about ABC Vacuum Cleaners. We're having a sale today for all callers who answer the following questionnaire... To receive our free coupon for vacuum cleaner bags with your fist purchase, just type in your fax number now..."

The messages can also include information about current specials, events, or directions to a store for example. In some cases, a caller can be automatically routed to the business in question, or can leave a message for the business if it is after business hours. These messages can be picked-up by the business so that the reader can be contacted the next day.

Operator Services

Operator Services systems assist live operators in the function of helping callers to "navigate" themselves through a telephone call. The "service" in the term Operator Services is the assistance in this navigation of a telephone call or attempted telephone call.

The item being paid for is the telephone call itself. This is in contrast to a telemarketing type of transaction, where the agent is taking orders for airline tickets. In the case of ticket sales or newspaper subscriptions, the transaction itself is the value-added service. Operator Services applications include Pay Station Message Forwarding, Calling Card Voice Prompting, and Collect Calls. An example of Partially Automated Directory Assistance and Call Completion is detailed below.

The environments where Operator Services applications are implemented are typically network-based. Companies like IBM, Computer Consoles, and AT&T install and service systems of this kind. Most of this operator services systems are installed in large call centers at the Class 4 Tandem level at Telephone Company Operator Service Centers.

Example of a typical Operator Services application: Directory Assistance

RBOCs (Regional Bell Operating Companies) have a high incentive to streamline costs in their Operator Services Centers. A somewhat folkloric calculation has been much touted by RBOC executives: "that for every second of time that is saved in each call for a bank of operators at any service center, $1,000,000 is saved each year."

With as many as 3,000 operators at each RBOC, there's a lot of money to be saved. And in addition to saving money, the RBOCs wish to generate revenue by providing premium services when possible. Without some level of automation, this would be virtually impossible to achieve.

Today, virtually every Operator Service center is now equipped with some form of automation. In most cases, a live operator answers a call for directory assistance and locates a listing by typing commands on a terminal which is connected to a centralized SIDB and LIDB (Subscriber Information Database and Line Information Database).

The SIDB and LIDB listings are cross-referenced to a telephone number which is then automatically sent to a voice response unit which plays the digits out to the caller, who is transferred to the unit automatically. The voice response unit also allows the caller to have the number repeated, or in some cases will route the caller back to an operator for further assistance.

More sophisticated systems will automatically collect the caller's ANI (Automatic Number Identification) in order to bill the caller a certain premium if they want to be connected to the number they asked for automatically. This is referred to as *Call Completion.*

Telemarketing

Both inbound call centers, and call centers making outbound calls are involved in telemarketing. In fact, with fierce competition in the field of telemarketing, almost every agency will provide both functions to a client.

Classic inbound telemarketing applications include ticket sales, airline and travel reservation systems, and catalog sales. Newspaper subscription solicitations, debt collection over the phone, and mass outbound notification are examples of outbound functions that sometimes require predictive dialing technology.

Voice Processing solutions are diverse in this application segment, because gear can be used solely for call processing (outbound dialers) in order to queue-up calls for live operators, or it can be used to augment the actual order-taking process.

Most Telemarketing solutions are found in CPE environments, although there may be growth in the network area due to the roll-out of some Centrex-Based ACD packages which could use voice processing capabilities. Customer premise ACDs are the most popular environment for voice processing solutions. ACDs use VRUs for telemarketing applications both with and without Host/PBX (CTI).

Example of a typical Telemarketing application: Newspaper Subscriptions

25 to 30 percent of an agent's time can be taken-up by simply placing calls, before a prospect is even spoken to. Each call can take from 30 to 45 seconds before it is determined whether or not the call can be completed, or whether or not the call is being made to an answering machine. This makes the cost of waiting to talk to a prospect very high for outbound telemarketing professionals.

The most obvious answer to this prob!em would be totally automate the function, however, except for limited uses, this has been rendered illegal by the Telephone Consumer Protection Act.

Now, outbound telemarketers must have a live person on the phone to speak with prospects. Because of this law, these calls are made manually, one at a time by each telemarketer.

Thanks to special call progress capabilities of today's call processing systems, the call center has been upgraded to a much higher level of sophistication. Sometimes called power dialers, or predictive dialers, these special systems make calls in a totally automated fashion from a pre-determined list that is stored in a computer. These calls are placed and then switched-over to an agent only when it is determined that the called party has answered the phone.

Agents are then able to spend their time speaking with prospects rather than dialing the phone numbers and waiting for someone to answer much of the time. Many of these systems will "pace" the rate at which calls are placed in order to keep up or slow down according to how long the agents are taking to finish each transaction.

Multi-Applications Platform

Multi-applications platform (MAP) systems weave together a variety of communications solutions. A "MAP" might have a combination of E-mail, fax, and voice mail services all combined in order to interconnect once disparate mediums of communication. For example, a MAP user could access his or her system by calling from a pay phone or hotel phone to pick-up messages. The user may hear a prompt which indicates that there are 3 voice messages, 2 E-mail messages, and 1 fax waiting to be picked-up. This person may then listen to the voice mail in the usual way, listen to E-mail via text-to-speech, and then use a portable PC to extract faxes all in the same phone call. Messages could optionally be sent into a MAP system in the same fashion.

In network environments, telcos and IXCs have been planning for AIN (Advanced Intelligent Networks) that take different forms of communication into account. Tariffed services such as BRI (Basic Rate ISDN) and CLASS (Custom Local Access Signaling Services) like ANI, for example, allow for the mix of voice, video, and data.

MAP is not just multi-technologies, but rather, multiple applications that are accessible during the same transaction within the same platform. An audiotex system with fax capability does not necessarily qualify as MAP. This is because the existence of a fax card in the system may not in itself change the original intent of the call: Access to stored electronic-based information.

The platform would qualify as being a MAP if the platform allowed callers to get voice information, trade messages, and then perhaps order a product, for example.

Example of a MAP application: Automotive Repair Status & Messaging system

Car dealerships are installing new systems which combine messaging and IVR disciplines in order to provide a better grade of service to customers. What some dealerships are doing is providing a means to automate and make available information regarding the status of a repair.

Here's how it works: When you drop off your car for repair, a repair order number is given to you. You are then instructed to call back on a special telephone number and input the repair order number in order to authorize repairs, check on the status of your car, and to find out when it will be finishcd and how much it will cost. The dealership installs a call processing system that either has its own dedicated outside lines, or has it hooked-up to an extension on their internal telephone system. The call processing system is connected to an on-site repair processing system and to the invoicing system as well.

When your car's problem is diagnosed, the repair person picks up a phone and calls into the system to leave a message for you under your repair order number. This message indicates the scope of the repair and any unforeseen costs. You can then call into the system and type-in your number. You will then be prompted to type digits to indicate that you authorize the repair. Subsequent calls into the system will provide you with information on the status of the repair and costs.

More sophisticated applications may allow for the automatic entry of your credit card number to pay for the service, and an automatic outbound call to your phone number to alert you that the car is ready to be picked-up. This combines the functions of a voice messaging system and an interactive voice response system, so it would qualify as a true MAP implementation.

Audiotex

Audiotex applications provide for the basic retrieval of information that is heard over the phone. In many cases, the information is supplied in a passive fashion, that is just played-out with no action taken by the caller excepting the initiating call itself. Many other applications are more sophisticated, requiring the caller to select from a menu of choices by dialing digits on his or her touch-tone pad. The kind of information supplied includes everything from sports, weather, product pitches, to entertainment. The entertainment category includes polling and conferencing capabilities as well. Some audiotex information is provided at no charge, which is the case in most corporate-sponsored systems. A premium rate is charged for Audiotex messages that are dialed over 976/900 type lines.

A system that provides a two-way communication, such as those that cause changes in the database of a remote computer (see IVR), use an audiotex-like function to provide information on how to use the system.

The key in having a "pure" audiotex system is that the information being conveyed is the intent of the call, or is the medium itself. Audiotex applications may require an outboard host or foreign computer link in order to provide the information it "speaks" out, but the intent is to provide a unidirectional information flow to the caller.

Audiotex applications can be found both in CPE and Network implementations. CPE Audiotex systems are either stand-alone, or behind a PBX. Network environments for Audiotex are evidenced by "gateway" systems that are available through the major long distance carriers.

Example of a typical Audiotex application: Concert Schedules

Community bulletin boards, concert and event schedules are often sponsored by local radio stations, convention centers, theaters , and civic organizations. In the past, these services have been offered by loading pre-recorded information onto a bank of answering machines. This is now an archaic and often expensive proposition, requiring callers to listen to the entire tape in order to find the desired information.

An interactive call processing system can solve many of these problems. First, a centralized system can have many telephone lines installed, which all access the same "tape." In the case of a call processing system, the "tape" is a computer disk that can provide dynamic and non-sequential access to the desired information.

Many sponsoring establishments of such systems can provide security codes to authorized persons so that information can be updated remotely and on a timely basis. This provides for not only current information, but also a more interesting dialogue, because the personality and voice of each sponsoring establishment can be different.

Network-Based Voice Processing Solutions

Network-Based Solution Provider Requirements

Companies that provide network-based services are sometimes referred to as enhanced service providers (ESPs). A typical ESP is a telephone company like NYNEX, Pacific Telesis, or NTT. Long Distance carriers like AT&T, MCI, and Sprint also provide network-based services ranging from voice messaging, audiotex services, and fax gateways. There are a number of critical items on an ESPs voice processing checklist. For example, ESPs need systems with a large "Single System Image."

This means that even though hundreds or thousands of lines may be deployed, that it is likely that more than one system will have to be networked together somehow to get the job done. A "Single System Image" ensures that the voice processing array can be controlled, maintained, and accessed as if it were a single unit.

ESPs require a certain robustness in their services platforms as well. Systems that are installed in central office environments need to pass both (LATA Switching Systems Generic Requirements) specifications. The LSSGR defines how telco-grade rack-mounted platforms must connect to wiring harnesses, central office DC power, and perform diagnostic, alarming, and remote maintenance functions. Most of these issues have been demystified by Dialogic, who supplies LSSGR-compliant computers to ESPs which house voice processing gear.

ESPs also prefer to purchase gear which conforms to a Standards-Based Architecture. Virtually every large telephone company now has a set of rigid criterion for the ongoing purchase of computer-based technology. These criteria include the demand for sourcing equipment that is flexible, scaleable, & extensible. This mentality has been formulated in order to have better control of the many vendors who sell computer equipment to the telcos, and to provide telcos the ability to "swap" pieces and parts of computer systems to achieve economies of scale and take advantage of favorable pricing and vendor competition. To date, switch manufacturers have been able to escape these rules, because they have not yet conformed to any industry standards for interoperability.

A standards-based architecture also provides ESPs with a rapid means to access new technology from the best vendors. This is important for the ESPs, because they can upgrade their services quickly and respond to market needs faster than they have in the past. Because ESPs depend on mission-critical reliability, they tend to lean towards architectures which support fault tolerant host computer control of peripherals such as voice processing technology. Today, there is a mix of equipment that is being used for enhanced services that is both "Self-Hosted" and "Remote Hosted."

Technology Landscape

Network-based service providers are beginning to incorporate technologies such as facsimile, voice recognition, and speech synthesis. The platforms that they use to provide enhanced services must be designed in such a way that it is easy to add these technologies. ESPs don't want to re-creating the wheel by inventing new interfaces for facsimile, or text-to-speech, for example. They feel more comfortable knowing that there is a standard driver structure, bus structure, and line interface scheme to add these things on. The confluence of these technologies is in itself creating demand for architecture standards, and this is happening as a result of market demand in other sectors, as well.

Market Issues

Enhanced Service Providers, and especially telephone companies are experiencing a shift towards a profit center mentality versus a cost center mentality in their offerings. For example, the operator services center at a telephone company has long been regarded as a resource drain for telephone companies, and a huge expense. Over the past ten years, telephone companies have introduced premium services at the operator services center that are revenue-generating. For example, they are providing call completion on directory assistance calls and specialized intercept services.

As consumers demand higher grades of service and more service options, the ESP's will have to accommodate the demand by incorporating more and newer technologies. Although centralized development of enhanced service applications has been popular with telephone companies, they are realizing that new offerings require both internal and external development.

Examples of Network-based voice processing solutions

Figure (COMESS.DRW) diagrams the way voice messaging services are provided by telephone companies, cellular providers, and long distance providers. To a certain extent, the voice messaging system is designed much like a CPE-based system would be except that it's much larger, and has to connect to a variety of other gear.

Perhaps the most important concept is the idea of aggregating high amounts of traffic from as many as five to hundreds of end-offices. The network-based voice mail system has to be installed at the "access tandem" level of the switching system hierarchy in order to service a large area. The actual telephone traffic is routed through a large switch, called a tandem switch. The tandem then presents only voice messaging service calls to the voice processing system. Usually, the calls are routed this way as a result of call forwarding or a ring-no-answer feature at each end office.

In addition to the actual telephone calls, other data is collected from the end offices that identifies the nature of each call. This data is typically collected on a separate data link, called SMSI or SMDI (Simplified Message Desk/System Interface). SMSI links are multiplexed from each of the central offices and then presented the host computer in charge of the voice processing equipment. This allows the host to command the voice processing subsystem to automatically play out custom greetings to callers that are in the voice of the person who the call was intended for.

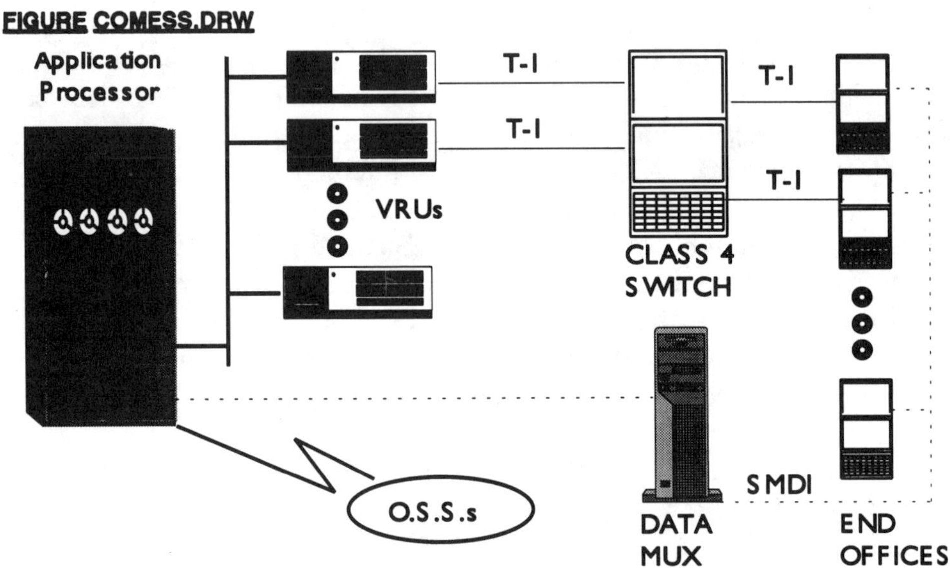

Figure (COINTER.DRW) shows how Enhanced Intercept Treatment is provide by telephone companies. The actual topology is very similar to the network-based voice messaging system. What's added is data links to a variety of databases which reside on other OSSs (operational support systems). These include the LIDB (line information database) and SIDB (Subscriber Information Database). The LIDB and SIDBs contain information about the rules for "intercepting" (re-routing) a call. For example, these databases include codes for whether or not a subscriber has paid their bill, asked for service to be disconnected, or if the line is temporarily out of order.

When a call is made to a disconnected number, for example, the call may be intercepted and connected to a voice response system. The digits including information about the called party are collected by the voice processing system either in-band (as part of the actual telephone call), out-of band (as with ISDN), or over a separate data channel.

FIGURE COINTER.DRW

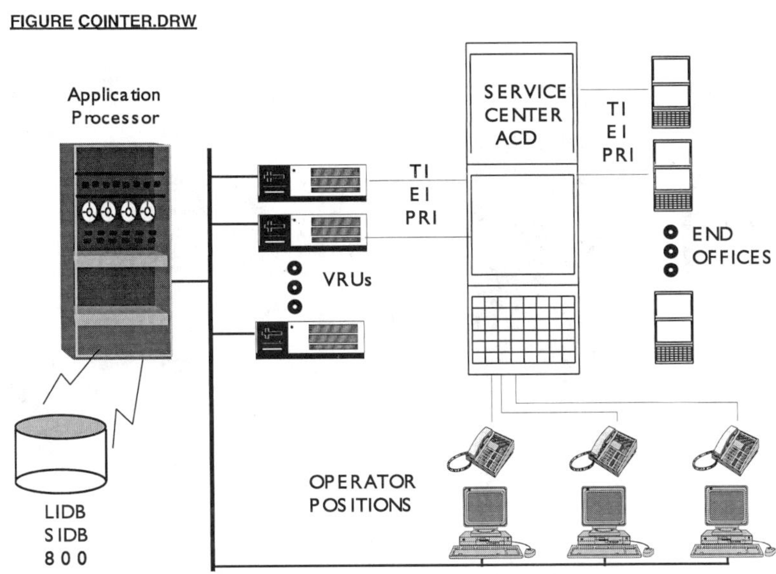

The host which controls the voice processing equipment will then do a query to the LIDB and SIDB in order to know how to "treat" the call. If the host finds out that the person has moved, and it's determined that a special recording is available, the host will locate the message number in its own database, and then command the voice processing system to play-out that recording to the caller. More often than not, the intercept treatment is a message like: "the number you have reached is no longer in service."

Figure (COSELF.DRW) helps to explain Custom Calling Self-Provisioning. Provisioning is the term telephone companies and other service providers use to explain the way services are "set-up" by accessing a variety of OSSs. For cxample, the BellCore-written MIZAR (not an acronym, just a name) system is a provisioning system for partially automating a central office switch for "recent change" information. Recent changes are the changes in class of service and features that are specific to a person's telephone line. For example, call forwarding is a feature that would require a recent change action on the switch's memory in order to enable the service.

FIGURE COSELF.DRW

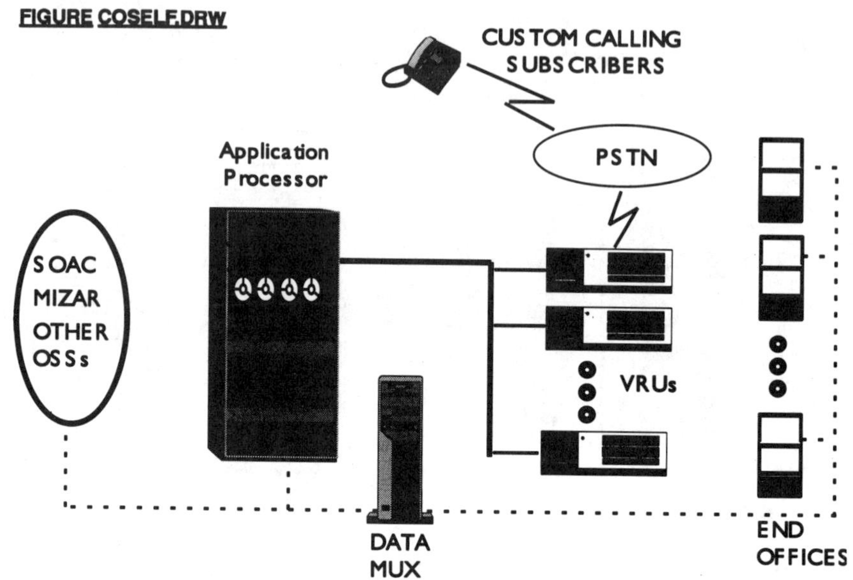

This provisioning is largely a manual process, requiring a call to be made to the service center, and a number of work orders to be written to provision the service. A self-provisioning system uses voice processing as a means to allow subscribers to enter the service order themselves. In some cases, the feature can be enabled this way almost immediately. This depends on whether or not the recent changes are batched or done throughout the day.

CPE-based Solutions Including IVR and Telemarketing

Requirements for successful IVR and Telemarketing Implementations

Although typically smaller, CPE-based voice processing systems can be as complex as their larger counterparts in the network. Makers of these systems also require a variety of resource modules and line interfaces.

While most network-based systems use either E-1 or T-1 digital line interfaces, CPE systems use both digital and analog lines including DID (Direct Inward Dial), E&M, loop start, and ground start lines. It's usually the density of a system and local telco service capabilities that dictate what type of lines are used, because a broad array of interfaces that conform to these protocols are already available from companies like Dialogic, Dianatel, and Execom.

CPE-based systems designers want tightly-coupled voice, switch and call processing modules so they can combine functions that are historically the domain of separate machines. For example, most call centers use separate host computers, voice processing machines, and PBXs or ACDs. Some vendors, like TRT, Cascade, and Teledirect are selling products which combine these classic functions in the same platform. An example of how this is done is explained below. What's important to remember is that in order for this confluence to occur, the builder of CPE voice and call processing solutions must have access to the latest voice, fax, recognition, and line interface technology from the best vendors. This is why open architecture is as important to the CPE players as with the Network players.

In addition to the hardware technology access, there are some core algorithms and detection sciences that are needed for many CPE-based systems. This is especially true of predictive dialing (outbound dialing) systems that are used in telemarketing scenarios. Predictive dialers automate the dialing of client telephone numbers and use "pacing" algorithms to keep a steady flow of answered calls available to the agents.

Answering machine detection plays an important role in predictive dialing systems, because they often "fake-out" many systems into thinking a real person is answering the phone. Since the purpose of predictive dialing is to present only true answered calls to the agent, the ability to determine the difference between a live person and an answering machine is a sticky issue. Companies like Melita and Dialogic are expert in these areas, so the technology is more readily available now than ever before.

The general need for call progress analysis is also a key buying criterion for systems integrators of CPE-based systems. This is true because CPE-based systems are often installed on the station side of key systems or PBXs. Since these systems are typically required to transfer calls to people who are using phones behind the telephone system, the call processing gear must be able to understand what a busy signal, ring, or "answer" sounds like.

This is especially difficult because despite the standards that BellCore (Bell Communications Research) has set forth regarding network call progress tones, most PBX, ACD, and Key/Hybrid manufacturers do not comply with these standards. There are literally hundreds of different telephone systems being manufactured each with their own set of rules on how to transfer, conference, ring, or indicate a busy signal.

This forces the need of great flexibility for the call processing system in CPE environments, because the software and firmware must be able to "characterize" the switching environment for the correct call progress tones in order to know how to process calls. This characterization can be done at system initialization by using special software that "polls" the switch. Programs like these will repeatedly call different extensions that are in a known busy or "no answer" or "answering" state. The resulting sounds that occur during these conditions are then analyzed by the software, so that the system can then predict what to do when these sounds are present while the actual application is running.

Call progress is less important when tight PBX Integrations are achieved. These tighter integrations are available either through telephone set emulation, or by using a separate data path between the call processing system and the switch itself. The data that is carried by these links provides the same information in most cases as does listening to the phone line for call progress tones. For example, the D/42-NS product marketed jointly by Northern Telecom and Dialogic actually emulates the functions of a Norstar proprietary telephone set inside of a PC which is the basis of a call processing system. The same chip set that is used in the Norstar phones are actually on the D/42-NS card, so the BRI (Basic Rate ISDN) data that the Norstar system uses to communicate with its telephones is available to the call processing application.

There are similar PBX integration products available for NEC, Mitel, and Rolm switches for example. This need for more extensive control over switches from the application is not limited to call processing machines. The general area of host computer to switch integration (called CTI, which stands for Computer/Telephony integration). Host to Switch integration is a hot topic for makers of CPE systems. At this point, most call center solutions require host processors to speak not only to call processing systems, but also over CTI links to switches. Maintaining these dual links is a burden for developers, and is one of the driving forces behind the SCSA-based ServerAPI. The ServerAPI combines call processing and switching commands over the same (out-of-band) command/reply channel in order to simplify these designs.

Technology Landscape

The technology landscape for CPE-based call processing is diverse. Systems integrators of these special platforms are faced with integration challenges that include the emulation of 5250 and 3270 terminals, LAN integration, and interoperability with a wide array of servers and computers. Buyers are no longer content with simple, stand-alone voice processing gear during the 90's. Their expectations have been raised by companies like Novell, Microsoft, and Sun Microsystems. The computing populace at large expects the telephone systems, call processing systems, and supporting technologies to work as well together as their LANs, Printers, Scanners, and PCs. Unfortunately, the telephone end of the business has not embraced standards and open architecture at the same pace as the computer industry. this is why SCSA is so important. SCSA helps to define cross-functional input and output points between these disparate systems.

Market Issues

Depending on the specific application segment, the year-to-year growth in CPE call processing is between 30 and 50 percent. With this growth, customers are demanding the integration of applications and systems. Today's buyer is beginning to ask the question: "If my file server is a PC, and my call processing system is a PC, and my fax server is a PC, why can't all the hardware be in the same box under the control of the same software. Or at least why can't they all talk to each other under the control of the same software?"

This very real question is fueling the emergence of MAP, and is causing the entire call processing industry to re-think how things are done. Customers are also now requiring integration with existing databases with minimal disruption to their existing computing environments. What this means is that the "shoot-em up and woolly" industry of call processing, where "anything goes" is beginning to mature and take notice of common interface and user interface requirements like their counterparts in the (regular) computer industry.

Examples of CPE-based Call Center Solutions

Figure (HITRAD.DRW) details a traditional IVR (Self-Hosted) application. The voice processing platform in this case is the software state controller for the application. It answers the phone, collects information from the caller, and then communicates with the host computer through a protocol conversion card or local area network. In essence, a caller is an extension of the terminal emulation function that is going on inside the IVR system. Each tap of the touch-tone phone pad by the caller, in effect, is the same as a keystroke by an agent who would be using a 3278 terminal.

FIGURE HITRAD.DRW

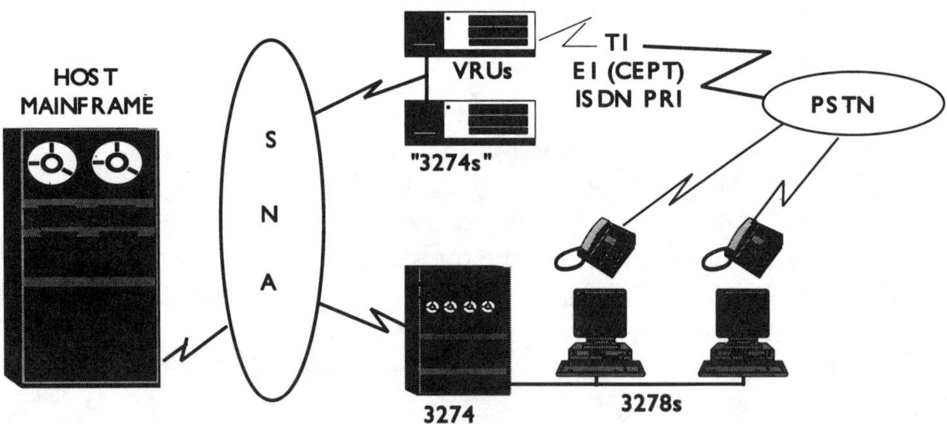

In many cases, the voice processing platform is acting in the role of multiple terminals and also a cluster controller, so that multiple cluster controller sessions can be carried on simultaneously. For example, an IVR platform with two cluster controller emulation cards can allow 64 concurrent SNA sessions with the right software installed. All of this is done in a very small amount of real estate (inside a PC or two), instead of with 64 desks, chairs, monitors, and terminals. The desks, chairs, monitors, and terminals are all "virtual," and replicated by the telephone caller him or herself.

Much like the previous example dealt with the emulation of a data terminal, PBX Integrations With VRUs do the same thing, except they sometimes emulate special telephone instruments instead of data terminals. This is diagrammed in Figure (PBX&VRU.DRW).

FIGURE PBX&VRU.DRW

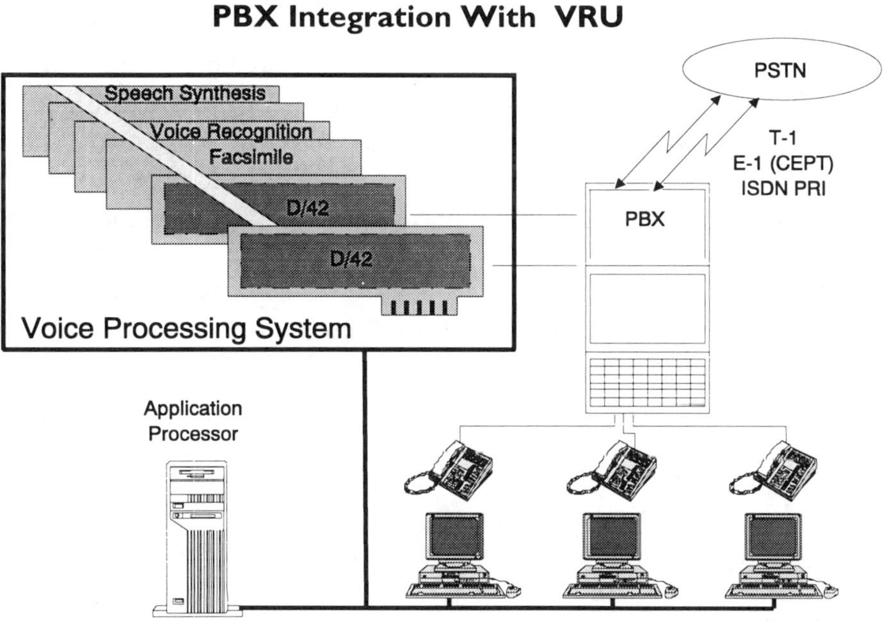

What's interesting is that nowadays, the digital telephone instrument which is used behind so many PBXs and Hybrids is looking and acting more like a data terminal all the time. For example, phones made by AT&T, Northern Telecom, and MITEL all have LCD displays, data ports, special "soft" function keys, and other capabilities that you would expect to find on a PC. Most of these functions are enabled by tight interaction with the switches' resources. These resources include conferencing circuits, data links, and access to telephone-based functions such as transfers, calling party identification digits, and so-on.

Some of this data is passed back and forth between the switch and the telephone by using COV (carrier over voice) circuits. COV-based systems include the MITEL PBXs, which send data in modem-like format above the (3000Hz ceiling on the human voice). This is all done over the same circuit used for phone extensions, so that voice and data can be carried simultaneously.

Some PBX manufacturers simply transmit the voice and data over separate lines to the telephone. Still others use their own version of basic rate ISDN, which is the case with certain Northern Telecom, and AT&T sets.

The biggest trend in convergence of the voice processing and switching technologies is putting all of the aforementioned technologies in the same box. Figure (CCBB0.DRW) diagrams the way this can be done in order to build small to medium-sized call center systems. In this example, a 48x24 Agent Configuration is pictured using off-the-shelf PC expansion cards. The cards needed to build such a system are readily available from Dialogic, Dianatel, and others. The line interfaces can be analog or digital trunk cards. For example, two DTI/211s (Dialogic T-1 interface cards) can be used as the trunk interfaces and several multi-port voice cards can be used to play prompts, record messages, and listen for digits and ANI (automatic number identification) routing data. Modular Station Interfaces can be used to connect the on-site telephones with the voice processing system.

In this example, the software to control the array of equipment is hosted by the same PC which is housing the software. The distinction between a "call center" system, which may be equipped with predictive dialing versus a "plain old" corporate communications system is a matter of application software. This configuration could be used as a small PBX or hybrid key system, in fact, by using ADSI-based screen phones (Analog Display Services Interface), functions similar to proprietary PBX phones could be emulated. Telephone Response Technologies of Roseville, California manufactures platforms that use off-the-shelf components to build these center systems.

Systems such as this will become available over time which use the full power of the SCxbus (explained later in this book), so that the number of stations can easily reach 1,000 or more phones in the same system.

FIGURE CCBB0.DRW

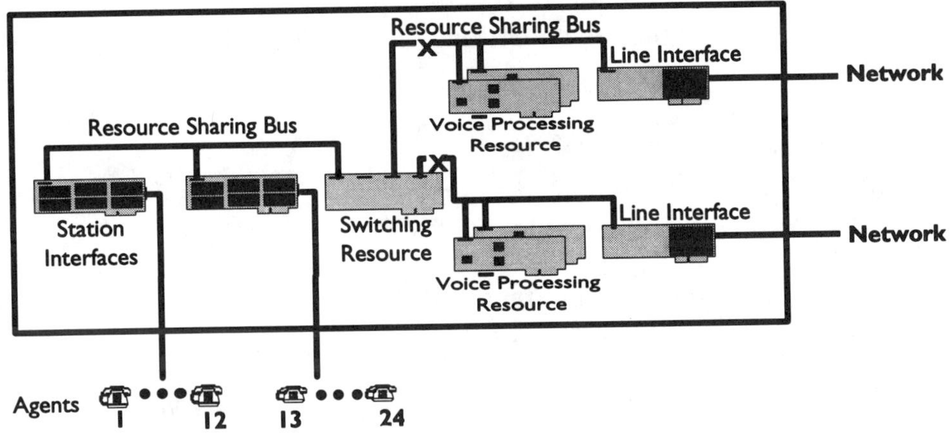

CHAPTER 3 - SCSA: An Overview.

This chapter provides a high-level overview of SCSA (Signal Computing System Architecture). SCSA is a cross-industry initiative that has over one hundred supporters including IBM, Tandem, NEC, Northern Telecom, and Siemens. It's also a very special initiative inside Dialogic, the company who spearheaded SCSA. The architecture will help to provide an infrastructure to bring the complex picture of computing, telephony, and voice processing together:

- Platform and Bus Standards
- Firmware Interfacing Standards
- Driver Development Input and Output Standards
- API Standards
- Alternate Computing Model Standards
- Out-Of-Box Networking Standards

SCSA represents not only some technological breakthroughs, but it also represents a development initiative for many new products. These new products span several industries and are being designed by many companies that you will may already be doing business with (and also competing with in some cases). SCSA is a set of specifications and an industry-wide development initiative. In other words, SCSA is not only a concept, but also explains the details of how new products will be implemented. Here's a snapshot of some of the challenges that SCSA tackles. These are explained in further detail in later chapters:

- The challenges of rapidly changing technology
- The challenge of standards unification in voice and call processing
- The lack of standards that go beyond the baseline of a switching bus and all the way up to APIs

The endorsers of SCSA believe that this new architecture will help us to embrace these challenges and most importantly, help customers to be more successful, which will in turn expand our marketplace and opportunities.

The ideas manifest by SCSA are the result of years of research and focused developments between Dialogic and other industry leaders. Although Dialogic is the main proponent of SCSA, it is important to note that the specifications come out of a group of industry experts and endorsers. Dialogic took special care to work with many companies, including competitors, to put this architecture together. SCSA is not manifest in one company, as ISA or EISA is not supported exclusively by IBM or Zenith.

Who is driving SCSA?

Dialogic is driving SCSA, but a variety of companies are designing to it. For example, when you review the discussion of the ServerAPI and its supporters, you'll see that both Tandem and ViCorp International made significant contributions to the draft standard circulated by Dialogic over the past several years. In addition, companies like Expert Systems, Voicetek, and Global Communications also had a hand in directing the outcome of the ServerAPI.

Even now, Microsoft's T/API specification is having an influence on the proposed adaptations to the ServerAPI for Phase III (client/server) roll-out. There's no doubt at this point anyway, that Dialogic has more "skin in the game" than some of the other contributors, but the company really wants the bus specs, driver standards, APIs, and the rest to be used and adapted into everyone's implementations.

When we get into the discussion on the SCbus Supervisory Channel, you'll see that it's detailed in such a way, that numerous vendors can comply with the bus standard without necessarily buying an HDLC chip from Siemens, for example.

Who's driving MVIP?

Natural Microsystems hosts the annual MVIP Developer's Conference, and they invite licensees talk about their bus standards. There is a core group of supporters who seem to provide the most direction for MVIP. This group includes Mitel, Natural Microsystems, PictureTel, GammaLink, and Voice Processing Corporation.

Mitel has a vested interest in supporting MVIP, because the bus standard is based on Mitel's STbus, which uses a Mitel semiconductor ship set. Mitel also has an interest in making STbus (MVIP) a success because they see it as a way to attract more applications development onto their PBXs.

SCSA MISSION

The supporters of the SCSA initiative want to enable industry growth and customer success by promoting industry standards for high performance call processing systems. SCSA constituents are trying to lower barriers for technological advances. In addition, supporters want to support an architecture that will handle advances in voice, data and image technology.

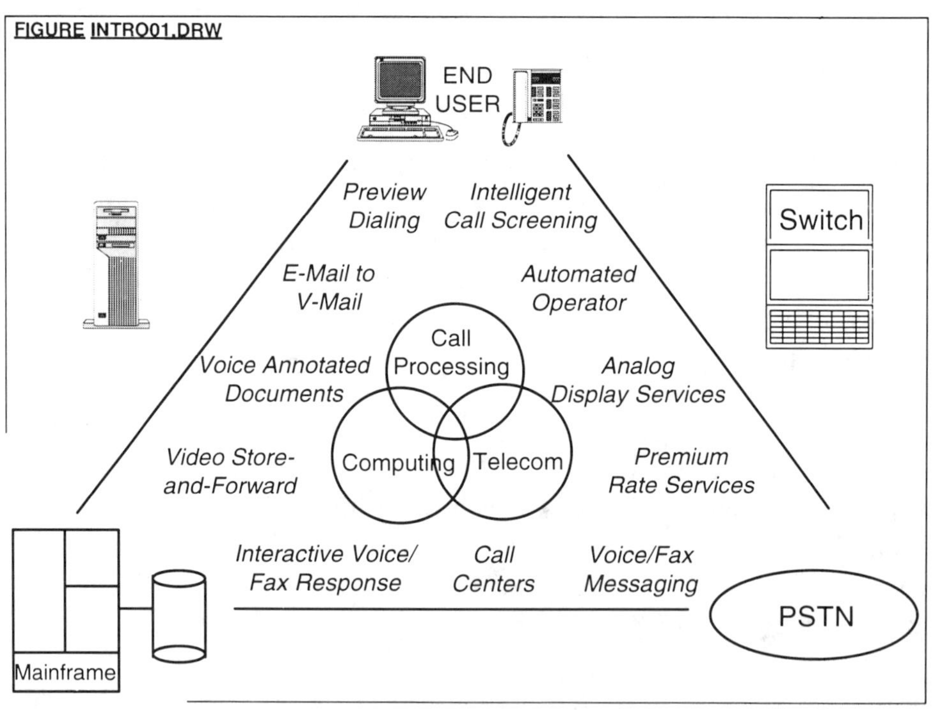

Call Processing represents a fulcrum between access to computers and access to telephony. This is because all processing basically automates telephone traffic and computer access at the same time. Figure **(INTRO01.DRW)** shows the relationship between the telephony elements and the computing elements in modern communications networks. On the telephony side, there's switches, line interfaces, and premium network services. On the computing side, you have application processors, mini-computers, and PC's. In the middle of these two worlds, there's a discipline called signal processing. Signal processing manipulates the telephone signals, the on-hook and off-hook messages, and information coming over the telephone line.

So if you're wondering what the SCSA really means when we say Signal Computing System Architecture, it's where the world of Computing and Telephony meet at the lowest level. ***Signal Computing*** is a borrowed term from Analog Devices, the integrated circuit manufacturer. Signal Computing seeks to define a structure and interface for how you manipulate signals, how computers can manipulate signals and process them. What SCSA endorsers want to do is bring Signal Computing to the next level of platform-wide implementation.

Call processing systems combine telephony and computing functions with the help of signal computing. This includes access to the telephone (including both analog and digital T-1 cards), PBX integration cards, voice processing, fax, speech recognition, and even video processors. The basic voice store and forward capability allow callers to speak into the phone so the system can digitize their voice and accept touch-tones as commands.

Quick History of PEB

It's important to understand the basic premise of the first and second generation call processing and resource sharing busses so you'll be able to understand why this new SCSA architecture and SCbus are meaningful. *The PEB was the first generation digital TDM bus and signaling protocol for call processing.* Time Division Multiplexing is a fairly straightforward concept. It's the idea of being able to send multiple signals that represent disparate communications and to squeeze them all into one digital communication line. A good analogy would be a subway line -- a line that is shared by multiple trains. Each "train" has cars that represent portions of the communication, all of which are assigned a set period of time when they can get on the same track (or highway) and use it to get to their destination.

What's actually happening is the assignment of multiple conversations onto *time slots*. These conversations are digitized and sent along a digital track or highway. It's all happening very fast, and all of the communications get put all back together on the other end. In a sense, the PEB treats communications from various devices as if they were like fast-moving trains on a shared track.

This is why the PEB is called a Digital Highway. Physically, it's a group of wires all together in a flat ribbon that connect the cards together; and then the cards are able to send information back and forth and share information. Many telephone lines coming in through a network resource can stripe across that bus and pick up information out of each of the cards.

In addition, a resource card can simply listen over the PEB. They can listen to one of the time slots and say: "*Oh, I heard a yes*" (If it's a voice recognition card, for example).

The PEB was introduced by Dialogic in 1988 as low cost, high performance architecture for interconnecting resource boards.

It was really the first mezzanine bus type of architecture that we know of for low cost platforms. The PEB allowed developers to put together systems that are an order of magnitude less expensive than some proprietary systems. The PEB specification was released in 1989 to technology suppliers for integrating fax, speech recognition and synthesis with voice processing.

Dialogic went out and worked closely with a variety of manufacturers of complimentary technology to build cards that worked together over the PEB. The company formed relationships with Promptus, Scott, Voice Processing Corporation, GammaLink, and others.

This is how Dialogic helped to accelerate the many offerings now available to Dialogic-based developers. The PEB specification was formally opened in 1991 due to market demand. This was spurred to a certain extent by Natural Microsystems' MVIP initiative. At this point, PEB-based systems have the largest installed base with the greatest variety of third party products available today. There are over 300,000 PEB-based ports out of 800,000+ ports shipped by Dialogic.

Quick history of Multi-Vendor Integration Protocol (MVIP)

MVIP was the second generation TDM bus and includes software conventions for call processing. It was introduced by Natural Microsystems, Mitel Semiconductor and five technology suppliers in 1991 as a proposed industry standard. Although some MVIP players are more ambitious than the specification itself, MVIP is just a bus specification, and includes no software driver kits, firmware code, or APIs.

The word "protocol" is a bit of a misnomer in the acronym, because, the MITEL ST-Bus (which is the basis for MVIP) is a PCM/TDM bus specification, not just a switching protocol. The MVIP constituency has made a valiant effort to establish software conventions, but in the limit, it's only a bus. This makes for a major departure from SCSA as an entire architecture...

MVIP specifications licensed by Natural Microsystems.

No licensee is allowed to proliferate the specification per the MVIP contract, so as much as it is referred to as being "open," it wasn't and isn't really all that open. MVIP licensees are forbidden to share the MVIP specification with non-licensees, so it's not in the public domain. Natural Microsystems uses a special licensing contract which allows them to use your name in public relations and marketing.

This means that upon executing the standard MVIP agreement you are agreeing to not share the knowledge with anyone else. At the same time you are agreeing not to share the information with anyone else, you are agreeing to have your company's name "shared" by the proponents of the architecture for marketing purposes. SCSA gets away from all of this by encouraging parties to share the specification freely. SCSA supporters sign a separate form to allow their name to be mentioned in marketing material for same.

MVIP Provides increased bus bandwidth and switching capability over PEB

This second generation bus does make some improvements over the first generation of the technology. It allowed people to switch as many as 128 full duplex time slots, whereas the PEB only went up to 32. Going beyond that with PEB requires either Dialogic's DMX or Dianatel's SmartSwitch, which can handle eight PEBs. The increased timeslot capability of MVIP represented a significant advance for high-density systems integration until the introduction of SCSA.

Growing installed base due to strong marketing efforts

Natural Microsystems has done a respectable job of proliferating the MVIP specification. There are a number of MVIP installations, although they pale in comparison to PEB-based installations. Nonetheless, SCSA embraces all of those developers who have used MVIP so that they can take advantage of the advances in SCSA. That's the beauty of SCSA that this book will reveal.

The only truly damaging thing that has come out of MVIP is that NMS never approached Dialogic in the beginning to ask them to support it, but rather kept it a secret from Dialogic until after it was announced. This caused great confusion in the industry, and has actually slowed-down the design process of many systems. In addition, Natural Microsystems paid no homage to the hundreds of thousands of ports that were already out there and in use on either PEB or AEB, therefore "abandoning" thousands of customers when they kicked-off the second generation architecture. In contrast, those companies involved with SCSA will embrace all existing bus architectures, and will work with competitors to help the industry grow. For the record, top management of Natural Microsystems was formally approached by executives at Dialogic before the public announcement of SCSA.

Architecture Goals

Dialogic has a highly interactive relationship with customers and other industry leaders. The company sponsors industry roundtables, user group meetings, development seminars, and works closely with other manufacturers to constantly upgrade technology. On the basis of their supporter's input, the following goals were developed:

Open call processing system architecture that meets customers' requirements and supports future growth.

Dialogic and other SCSA supporters know that if industry leaders can help users to be successful, then the manufacturers will also be successful. An open architecture that includes drivers, busses, APIs, and firmware provides us with an appropriate and powerful vehicle to reach this goal.

Widespread industry input from leading computer and switch manufacturers, voice processing suppliers and technology providers.

This means that SCSA supporters must keep a broad scope and work closely with industry leaders such as Microsoft, WordPerfect, IBM, Tandem, and others. This will help call processing companies to stay abreast of emerging computing models and adherence to API standards. These newer API standards will go beyond the scope of call processing as it is defined today. For example, some of the work that Novell is doing on voice-based NLMs (NetWare Loadable Modules), and Microsoft's T/API will help to redefine the way computers and telephone work together.

Electronic compatibility with existing bus standards, including PEB and ST-BUS.

Of paramount concern is to maintain a commitment to the people who have already made an investment in call processing technology. The idea is not to abandon all the people who made products based on PEB. Not everything is going to always be 100% fully compatible but the goal is to maintain a commitment to the PEB architecture and not to damage current product investments. The bottom line is that SCSA will help keep people in the industry happy with their existing investments as much as possible.

Support for advances in voice, data and image technology.
The architecture needs to anticipate technology changes and allow developers to plan for current as well as future success.

Customer Requirements

Now that we've discussed what SCSA endorsers want, let's visit the direct feedback SCSA supporters have gotten from customers -

Open Architecture based on existing and future industry standards.

Dialogic found that people wanted an open architecture that was based on existing standards and some proposed standards. Customers want increased bandwidth, and more technology that interoperates easily.

Bus with increased bandwidth, switching and signaling capability to support high densities and emerging technologies.

They want the ability to add many more cards in the system and increase the amount of traffic that can flow through the system that can all be packed into one box. They do not want to be forced into second and third boxes unnecessarily.

Access to wide range of signal processing algorithms and network interfaces.

Customers want to pick and chose the best algorithms available, and would like to use them in systems that hook-up to worldwide telephone networks. Network interfaces need to be commoditized, such that parameters for making a system work in different countries are made simple and abstracted from the application program.

Customers also need the ability to buy not only software algorithms for voice recognition or text-to-speech from different companies, but also to work with many call processing resources from a variety of vendors. That's called the interoperability factor. A successful call processing implementation may require technologies from eight different vendors, and they all need to work together as seamlessly as possible.

Multiple documented interfaces to allow interoperability between call processing resources from multiple vendors.

The SCSA documentation must provide details on how to interface to every element of the architecture, and not just at the bus level. An SCSA-based platform includes hardware, firmware, device driver interfaces, and application programming interfaces. A general book of documentation to these access points is required.

Protection of software investment through APIs that insulate application from underlying technology and enable growth from small to large systems.

Customers would like to develop applications that will run in Self-Hosted, Host-Slave, and Client/Server environments. In order to protect the overall durability of a client's software investment, the solutions that are built should be able to work in a variety of environments. In addition developers would like these solutions to ultimately work with any operating system. This also speaks to the issue of host-independence.

For example, in the host-slave view, you have a mainframe or minicomputer where the application lives and all the control threads come out of it into voice response units with voice processing technology in them. With the client/server model you have a file server on a local area network, and everyone can access it to share things on the file server over the local area network. In this case, everyone has their own PC running some kind of client or "access" software.

This is where the concept of a voice server comes in -- where there's a PC housing an array of call processing cards. The "client" PCs don't have their own cards that do these functions, but every time someone has to send a fax or send a voice message to someone else, they access the voice server through the local area network and send commands back and forth to allow transactions to occur from the server (shared) standpoint. This makes for excellent cost/performance ratios.

Portability across processing system environments means that if an application is written in conformance to certain API standards, that it may work in a self- hosted, host-slave, or client/server implementation. Not only will these APIs insulate application developers from the underlying technology, but it will also insulate the programmer from the computing model that's being used.

A customer should also be able to build a 4-port application, and then be able to drag the application across a multi-node system. This means that applications built for small use should work in large environments and vice-versa. This is a big problem in the industry today. For example, large proprietary vendors must sell two or three different implementations of their voice mail or IVR offerings (all based on different architectures) in order to provide customers with small, medium, and large systems. Scalability improves inventory, sparing, training, and documentation economies as well.

Signal Computing System Architecture

FIGURE STAND0.DRW

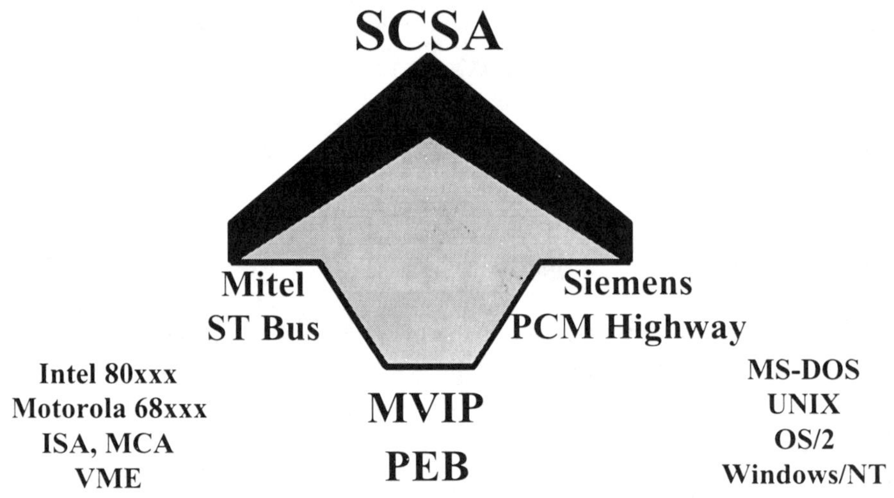

Figure **(STAND0.DRW)** really brings the intent of SCSA into perspective. Perhaps the most important thing to understand about this architecture is that it combines and extends the capabilities of products built to older standards under the umbrella of SCSA.

Recall figure **(INTRO01.DRW)** at the beginning of the chapter that spoke to the issue of "***confluence of technologies***": The computer processing side on the left and the telephony world on the right. Call processing was right in the middle, acting as the catalyst for the confluence -- making those two worlds come together.

Call processing is undergoing its own technology confluence -- wherein voice recognition, text-to-speech, voice store and forward, facsimile, are all coming together as well. This picture shows how *"SCSA represents the confluence of standards."* ...SCSA is to standards as call processing to disparate technologies. SCSA brings together many different standards in order to make it easier and quicker for a call processing system to be designed and successfully deployed.

For example, there's the Mitel ST bus (including MVIP), Siemens PCM highway, and PEB. There's four standards. This third-generation SCSA architecture provides electronic compatibility for all of these previous standards.

You could say that SCSA is the Standard of Standards.

From an imbedded investment standpoint, we're recognizing all of the hard work developers and manufacturers have put into designs based on these disparate standards. The point here is that we're not throwing anybody else's investment aside.

If somebody has invested in another bus or designed to another platform, they'll be able to incorporate that design within this new one to an extent, but that's not to say that they'll be able to do everything that you can do in a pure SCSA mode. That's the whole point of backwards compatibility. It's not to bring the old stuff up to date, it's to make the old stuff work in some reasonable way with the new stuff.

So if you want a formal description of SCSA, you could call it

> "***A Set Of Standard Interfaces, Conventions, And Protocols That Enhance Multi-Technology Call Processing System Design And Performance.***"

Standards don't mean the same thing to everybody. The word "*standard*" is almost a misnomer cause there's a multitude of so-called standards. Having hundreds of standards is actually the lack of a standard. Access to stable, well documented information (so that you can ***establish predictability***) is what standards are all about.

From a workability standpoint, standards mean predictability of design. Now there's many different standards, but as long as they're well documented, it's more predictable for developers to get the end result they are looking for.

Signal Computing System Architecture Model

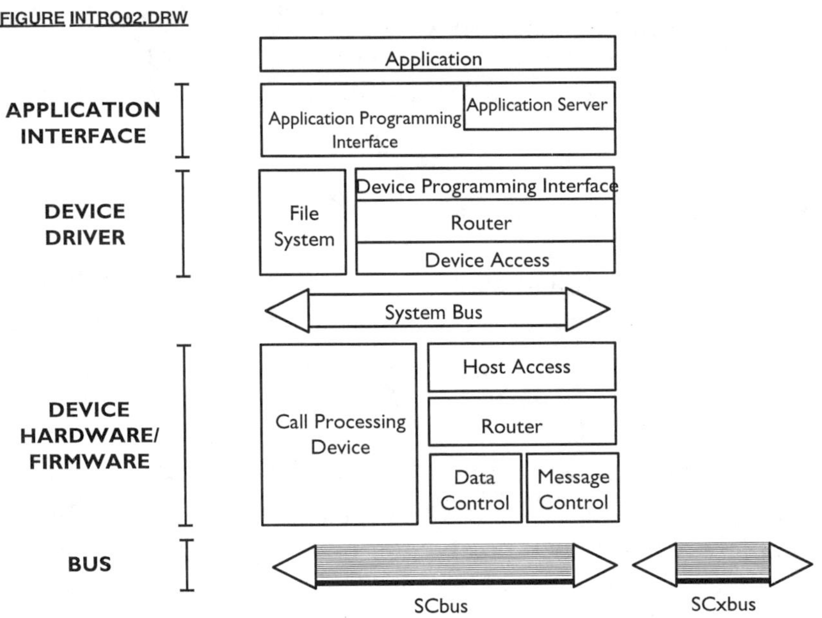

Figure **(INTRO02.DRW)** provides a snapshot of the entire architecture and really helps you to draw the distinction between SCSA as an entire architecture versus MVIP, which for all intents and purposes is just a bus specification. SCSA transcends the mundane bus "speeds and feeds" argument by encouraging standards for APIs and firmware hooks. In addition, SCSA encourages this at many different levels of this continuum. For example, SCSA establishes conventions for the bus level, resource level, supervisory and switching levels, and also for device drivers.

These are all graduated levels as you move up this (OSI model-like) chain of a call processing implementation. AS you can see, this architecture addresses all of the key components that are needed to build a call processing platform. And that is a key distinguishing factor between this and just a simple TDM bus specification like PEB or MVIP.

Quick snapshot of SCSA elements

Here's a snapshot of the primary elements of the SCSA architecture that we'll be going into detail on in the chapters of the book:

SCbus - Third generation TDM bus for transporting data and signaling between system resources.

The SCbus is much more sophisticated than the first and second generation buses that we are dealing with today. The SCbus is an open specification for serial bus which transports asynchronous data and message-based signaling information. The SCbus specification is being proliferated as an open standard. The strategy is to have the specification implemented in multi-sourced products from integrated circuit, switch, and computer manufacturers.

SCbus Switching and Data Handler - Integrated circuit for switching data between system resources and SCbus.

This element is manifest in an ASIC (SC 2000) which includes the SCbus data interface and switching handler functions. In addition, this is the element of the architecture that provides compatibility modes for PEB and MVIP. The SC 2000 ASIC will be used in all Dialogic-based SCSA product implementations. The strategy is to have SC 2000s available from parties other than Dialogic, such as VLSI or Arrow, for example.

It's important to note that while the ASIC design makes it easier to hook-up to the SCbus, specifications for the circuit will be made available such that developers can make products that are SCbus compliant without having to use the SC 2000.

SCmessage and HDLC Supervisory Channel - firmware and protocol for signaling between system resources.

SCmessage is the supervisory firmware element of SCSA. It provides for protocols for signaling and inter-module message passing capability. This will be used in all SCSA-based products which require signaling to be transmitted to or received from the bus. The idea is to implement the supervisory channel with a standard multi-sourced serial communications controller and HDLC protocol chip which is available from Siemens Components.

SCdpi - Common device driver services, message passing functions and device independent interface to system resources.

The device driver element defines the input and output protocols for message passing and communication between the APIs and the call processing system resources. SCSA endorsers already have a leg up in this area because of the de facto acceptance of the common driver services that are already in use. The idea is to provide standard interfaces to the generic device driver which eliminate the need to build a custom device driver for every new device which is plugged into the architecture. The generic device driver provides many system functions, such as queuing, scheduling, bulk data transfer, and diagnostic functions.

SCapi - Standard application programming interfaces to system resources and features.

SCapi is a collection of common application interfaces that have a cohesive look and feel and that hook into the top of the driver. These are these high-level libraries that provide the capability for a programmer to be abstracted from the intricacies and differences between separate call processing resources.

A good example of how this works is the way in which voice recognition cards from Scott, VCS, and VPC can all be manipulated in the same way from a developer at the API level, where commands to these cards are the same except for functions that are peculiar to certain advanced features.

SC *Server API - High level programming interface that enables client or host resident application to control a call processing server.*

The Server API represents a common set of high-level command-reply protocols and conventions which allow application software to be host-independent. This is the element of the SCSA architecture which fulfills the promise of portability across a variety of computing platforms. For example, if an application is written on top of this Server API, the application can be self-hosted, or in a minicomputer, or in a client/server implementation. This is the icing on the cake of SCSA - the ability to take a developer's software and to make it commercially viable in virtually any environment.

No you can see what the endorsers of SCSA mean when they say that SCSA is not just a bus - It's an entire architecture that provides a development environment for the most sophisticated platforms of the future.

SCbus

FIGURE INTRO03.DRW

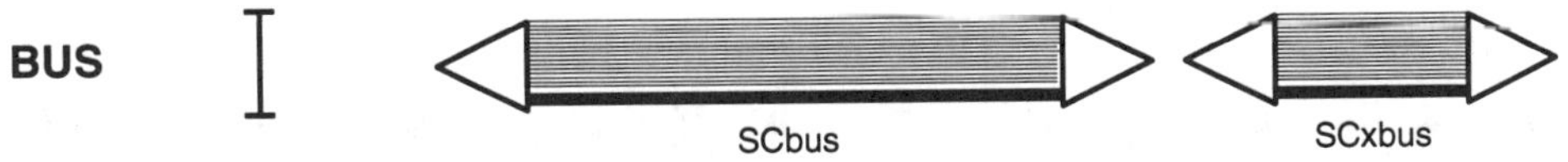

Figure **(INTRO03.DRW)** is a blow-up of the SCSA architecture snapshot which concentrates on the SCbus itself. This is the bus where all data, telephone, video, fax, and other traffic flows inside the call processing platform. It's a mezzanine bus that connects all of the Call processing resources together, and it uses the SCbus Switching Handler and the HDLC Supervisory Channel to control all of the traffic.

SCbus Characteristics

Open electrical / mechanical specification. No special license or contract will be required. The bus specifications will be in the public domain.

16 synchronous data lines. The bus itself will handle 16 data lines, which are supervised via a common signaling channel. This is to a certain extent, an in-box method of implementing an ***SS7*** (signaling system seven) type of model, where various transport channels are controlled by a data channel which acts as a traffic cop. This is a much more modern approach to building sophisticated and efficient call processing platforms.

Bus rates of 2, 4 or 8 MHz (4 MHz standard). The standard bus rate for the SCbus is 4.096 MHz, although the bus can handle bandwidth twice as wide. So as you can see, there's built-in growth to this bus in anticipation of greater user needs in the future.

512, 1024 or 2048 universal time slots (1024 standard). What's immediately apparent to developers about the capabilities of the SCbus is how it makes a quantum leap in terms of its ability to handle many more timeslots than any currently available bus in our industry. The number of time slots being implemented in standard mode is 1024. What that means is that over 1000 high-speed communications can be occurring at the same time on the bus. Now, if you want the conversations or communications to be real time between two people or full duplex, for example, that equates to 512 full duplex channels. And none of these channels are wasted on signaling or "unused" as with past generation busses.

When it comes down to just what you can plug-in from a standard telephony standpoint, the SCbus allows as many as 16 T-1 spans or 16 E-1 spans to be carried all in the same box. An E-1 span carries 30 discrete communications. So if you take 16 times 30, you can see how the math works out to a very robust communication highway.

Now, in comparison to the MVIP bus, for example, which touts 512 time slots, keep in mind that that's really only 512 (half-duplex) time slots which yields 128 full-duplex communication channels considering how timeslots are "stolen" in order to provide signaling.

Up to 130 Mbps synchronous data throughput (64 Mbps Standard). The effective throughput of the bus in its maximum mode is over 100 Mbps. Now that's the kind of throughput necessary to carry some serious traffic for high-density, mission-critical applications. It is now possible, because of this design to build large call center systems which mix voice, data, video, and other traffic for enhanced network services that many of our customers have been banking their futures on.

Separate 2.048 serial line for signaling. The SCbus specification includes an out-of-band signaling capability using HDLC protocol. This allows developers to take advantage of a recognized standard in signaling and apply it with off-the-shelf components.

Electrically compatible with PEB, Mitel ST Bus and Siemens PCM Highway. The bus is also electrically compatible with PEB, Mitel STbus (including MVIP) and the Siemens PCM highway. So here we have a bus, in terms of *speeds and feeds*, that's really superior to anything that's out there for efficient call processing.

SCbus Benefits

In terms of some concrete benefits for embracing the SCbus in call processing system design, here are a few points that developers are excited about:

> *Increased bus speed provides more time slots for higher density systems means more time slots for collateral technologies and higher densities.*

Most customers who wish to create high-density systems are in dire need of a higher speed bus, so that they are not forced into multiple platform topologies, or forced to go with some short-lived proprietary solution. The SCbus solves that problem. The bus is worthy of use not only in PCs, but also in tandem with large-scale computing platforms that can process thousands of transactions per second - such as OLTP, or on-line transaction processing systems. In addition, developers need to have access to multiple technologies per call. The SCbus allows developers to mix and match technologies in any ratio they desire. For example, a four-port fax card, a 16-port voice recognition card, and a two-port Text-To-Speech card can be shared across let's say 4 T-Spans. Or, a one-to-one ratio can be applied to the same system if this is required.

Simple and powerful timeslot control including non-blocked switching and broadcast between all time slots

This time slot control allows also for an optimal switching and broadcast mode capability where you can broadcast one time slot over all the rest of them for conferencing or other functions. This is a superior implementation over previous generations, because broadcasts are available over the entire spectrum of 1024 timeslots.

Full frame buffering and time slot synchronization, so time slots can be bundled for clear channel data transport

One of the most distinctive advantages of building platforms based on the SCbus is this timeslot bundling capability. Remember that with a 64 kbps timeslot, that's bandwidth of a regular telephone line; that's about the bandwidth required for a voice transmission or a fax transmission. But what about non-compacted full-motion, full color, real-time video? What about Bulk Data Transfer? The idea here is that you have the ability to bundle these individual timeslots together on the fly, so that your system can accommodate any kind of high-speed transmission that is required.

This enables a whole new age of microprocessor-based call processing. That's the whole idea with the architecture. It doesn't matter what you're sending across the pipe. As long as your technology conforms to the protocols and takes advantage of the Supervisory channel, you can set up and tear down clear channel and bundled communications at will. Now that's something that is simply not addressed by any of the current first and second generation busses.

SCxbus Out-of-box expansion capability for multi-node systems and connection to other computers and switches

In addition to the high bandwidth supplied by the SCbus inside of one platform, an expansion capability called SCxbus allows inter-node traffic between co-located signal computing nodes. This allows for the design of multi-node systems that have a single-system view. In effect, developers will be able to build systems that can accommodate thousands of lines, and allow resource sharing in-between the nodes for special feature access. The capability to build system redundancy is also enabled because of this bus extension capability.

For example, let's say you build a node and fill it with call processing resources and perhaps 10 T-spans coming into it from the network. Let's say you have a second system with the same number of network interfaces. The SCcxbus enables these two nodes to talk together -- for a real-time transaction to go into the first one and travel out the back and scoot down the expansion bus and back into the second node. This way, a second telephone transaction to access the second node from the network is not needed.

Automatic clock fallback handles network glitches without service disruption

One of the most severe drawbacks of previous generation buses, is that clock fallback capability was either totally absent, or not very flexible from a programming standpoint. With the SCbus, a bad clock on a primary resource card can be automatically switched over to a secondary clock source. This allows for a higher degree of up-time for mission critical applications.

SCbus Data Control (SC 2000 ASIC)

Multi-sourced integrated circuit with data sheet and application note provides open, cost effective access to SCbus

One of the greatest advantages of the SC 2000 is that this integrated circuit saves real estate on resource boards, simplifies development, and will be commonly available from more than one vendor. In addition, the specification for how the switching handler works will be made commonly available, and in the public domain, so that developers can use discrete technology to emulate the Switching handler's functions if they chose to.

PEB and MVIP "compatibility modes" enable technology suppliers to implement SCbus, PEB, and MVIP products with single hardware design

There will be certain modes that the SCbus can be put into so it operates strictly as a PEB bus or an MVIP bus. This protects the durability of an existing hardware investment for research and development efforts that so many developers have built to.

Unfortunately, the high speed and capability of the SCbus needs to be throttled way back to support the MVIP mode, since we will be using not even half of the bus's capability. Even so, those companies who have built PEB products and built MVIP products, this will be good for them. They can come onto the advanced SCSA bandwagon and not be abandoned by this leading edge technology.

They're not going to be left to drown, because it's a good thing to be able to protect their investment. Many technology developers (folks either with PEB or MVIP connections) did the industry a good service in making their technology available on the previous generation busses, so the architecture should help them to always remain good "industry citizens" in promulgating a standard that supports every generation.

SCbus Message Control (HDLC Controller)

> *Multi-sourced HDLC controller chip (Siemens) with data sheet and application note provides open, cost effective mechanism for managing signaling*

As with the Switching and Data Handler, the SCbus Message and Supervisory Bus chip will be multi-sourced. In this fashion, developers can decide on what component suppliers they wish to work with can serve their needs. In addition, having a second source ensures a safer initial design in case of shortages from one manufacturer.

> *Separate common signaling channel provides ISDN like signaling to meet real time response requirements of network*

One of the biggest challenges in building large-scale systems is the fact that the telephone network applies fairly rigid response time constraints. The timing thresholds for answer supervision and recognition of transaction requests from the network are so demanding, that dependence on application software control is not optimal.

The SCbus Message Bus enables quick, inter-board response to network requests for instantaneous response. This abstracts a lot of the "grunt work" from the application and protects developers from building systems that put trunks "high and dry" due to slow response.

Message based protocol enables enhanced communication between system resources

Much like ISDN, and SS7, the message bus uses a message-based protocol for setting-up and tearing down transactions. This is significant, because it provides a "look ahead" type of capability which speeds the efficiency of transaction flow.

This message passing capability allows network inside the signal computing node to act as a sub-set of the telephone network. In this manner, SCSA enables peer-to-peer and advanced inter-board communications which extend beyond the single-node system and even into the multi-node systems described earlier.

SCmessage

Figure **(INTRO04.DRW)** provides a cut-away view of that portion of SCSA which concentrates on firmware at the Call Processing Resource level. This is the layer that sits directly on top of the SCbus and directly below the Host interface and the SC Driver.

FIGURE INTRO04.DRW

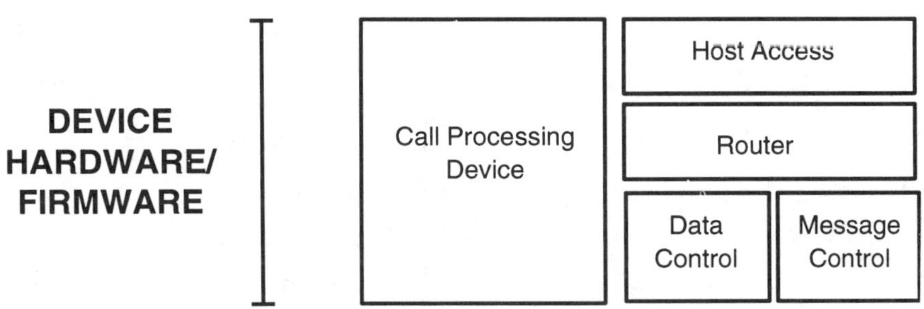

Open specification of switching commands and message protocol

This open specification allows Dialogic to work very closely with technology suppliers for technology transfer to speed development, and also allows developers to do things on their own without Dialogic being involved.

Firmware for managing switching & handler and message bus

Developers will be able to build source code variations for both Intel and Motorola processors for on-board control of the Integrated Circuit. This provides flexibility in being able to quickly implement SCbus designs on resource boards that use a variety of imbedded processor controls.

Layered software implementation of SCbus interface and host interface

This approach to resource board development simplifies both Dialogic's and other technology supplier development cycles. By documenting and providing specifications at several layers, developers of core technologies can concentrate on their distinctiveness in that core technology, and not have to worry about host interface and bus interface control.

SCdpi

This model provides a cut-away view of that portion of SCSA which concentrates on device programming interfaces in software. This is the layer that sits directly on top of the Call Processing Resources and directly below the actual Application Software itself (see figure **INTRO05.DRW**).

FIGURE INTRO05.DRW

DEVICE DRIVER

File System

Device Programming Interface

Router

Device Access

SCSA provides documentation on the access points to a device driver model. The device driver saves developers time in bolting-on their own call processing resources without having to re-do work already done by other manufacturers. A good example would be the Windows operating system. A windows developer who wants to put together a new word processing package, for example, does not have to write 100 printer drivers, because they are already handled by the Windows Shell. The SCdpi works similarly.

> *Software to provide common device driver services required by most call processing resources*

These common services include:

- Bulk data transfer, buffering and storage management
- Command / reply synchronization
- State maintenance

> *Device independent interface eliminates custom device drivers, speeding time to market and facilitating integration of technologies*

The idea is to eliminate the waste of energy, extra memory, and CPU drain of running multiple device drivers. Having a device independent driver enables quicker development because APIs become more ***Object Oriented***.

This speeds time to market because development time for independent drivers is virtually eliminated altogether. This helps both systems integrators, resellers, distributors, manufacturers, and end users alike.

SCapi

Most Dialogic Application Programming Interface specifications are already in the public domain, however, new ones will become available which will help to bring new technologies into the SCSA architecture. SCSA endorsers have formed a working group to provide robust additions to existing call processing API standards.

FIGURE INTRO06.DRW

APPLICATION INTERFACE

Application Programming Interface	Application Server

Figure **(INTRO06.DRW)** This graphic shows the multiple levels of interaction now possible with SCSA-based APIs. Note that a portion of the SCapi is divided to show the Server API, a high level programming interface that enables client or host resident application to control a call processing server. The specifications for these interfaces are detailed in the SCSA technical specification document.

Open specification of SCapi commands and responses

This open specification allows SCSA developers to speed their applications to market because a common API enables portability. The end result is that conformance to this highest level of the SCSA architecture provides absolute application portability, and absolutely sets SCSA apart from simple bus-only architectures.

SCapi provides object oriented interface and higher level of abstraction to call processing resources

These APIs are high-level libraries which abstract the application programmer from the intricacies of the underlying call processing technology. The goal in using these APIs is to make the application as device independent as possible.

Designed to interoperate with industry standard APIs

By conforming to standard APIs, interoperability is facilitated, so that applications will work with many different vendors' equipment. For example, hundreds of modems have been built to conform to the Hayes command set, which, in effect is an API. Manufacturers in call processing are also beginning to do the same thing.

SCapi insulates application from underlying technology and operating system

The SCapi allows developers to code applications on small development platforms, which can then be ported over to larger platforms, such as minicomputers and mainframes. This expands the marketplace opportunities for everyone who is involved in the call processing marketplace today, and allows developers to serve a greater audience of end users.

Some examples of conformance to common API structures include Rhetorex and PIKA, for example. Both Scott, VCS, and VPC conform to the same generic call processing resource API being promoted here. The idea is to broaden the scope of this initiative to include video, fax, and other technologies as well.

The end result of API standards is that application developers can plan their solutions more effectively, and also have the flexibility to chose the best component to do the job without the worry of driver conformance. This equates to quicker development cycles.

Server API enables client or host resident application control.

Native host can control an SCSA-based call processing server, allowing the application to be centralized in a mainframe or minicomputer, or distributed on a LAN. As discussed briefly before, the computing models that can be supported with the Server API include *Self-Hosted, Host-Slave*, and *Client/Server*. This is possible, because the API can reside inside of a PC, which houses call processing resources. Any developer who writes application software that uses the Call Processing API can then use the resulting code to control voice processing in a native host environment via asynchronous/RS-232, or TCP/IP means.

This high-level interface can be related to the CTI, or the Computer/Telephony Interface phenomenon. The problem with most CTI implementations is that they require a major dollar investment for licensing and generic software upgrades for switches and computers. Proposed CTI standards are really not OPEN, and they entirely skirt the issue of multiple technologies. Current CTI standards also presuppose that the application resides in a host computer that drives a switch. But what about fax and voice processing, and voice recognition, for example?

The Server API's byte-stream protocol includes commands for setting-up and tearing down calls, switching, voice processing, fax, and so on. It is an entirely open specification that allows any developer to build applications which control these multiple technologies today. This part of SCSA has gotten a kick-start by Tandem Computers and ViCorp International, who are using the Server API to control multiple Call Processing Systems from a non-stop minicomputer.

The next few chapters provide a more in-depth view of the technical aspects of the architecture, and also some applications and associated market opportunities that can be tackled with SCSA.

CHAPTER 4 - SCSA Applications

This chapter concentrates on some of the unique challenges we face in the communications industry, and how SCSA enables solutions to those challenges. First, we will discuss some of the Key Attributes of SCSA, and how these attributes can be leveraged. Next, I will touch on some of the elements required for advanced applications.

At the end of the chapter, some examples of SCSA-based applications and topologies will be reviewed. These examples are more advanced than those discussed in the first few chapters, so you may wish to review the earlier material for some background.

SCSA Key Attributes

In general, SCSA leverages existing standards and makes a quantum leap in speeds & feeds at the same time. For example, if you want to incorporate a new SCbus compatible card into a PEB-based system, the card will be "backwards compatible" with the other PEB-based cards in your system.

If you want to use the same card in a new system that has other SCbus components in it, then it will run at the higher speeds available to SCbus systems. This baseline compatibility issue is a good place to start when considering an architecture and when designing a new system or service. No one wants to be held hostage by yesterday's technology.

In the world of computation, SCSA supports varied form factors, processors, and host computers. This ability to support a wide array of control gear is an important discipline when building an advanced application. The reason why this is so critical is because from site to site and from customer to customer, the computing environment changes radically. It's a good thing to be able to change host processors and topologies quickly, so that you can adapt to whatever environment is the current one.

Scalability and extensibility is as important in call processing as it is in computation circles. For example, the ability to design single & multi-node systems allows applications to grow when needed, without necessarily doing a "forklift" upgrade. In addition, scaling the hardware implies that the applications portability issue is taken care of. What's the sense in having a scalable hardware platform if the software cannot be ported. SCSA encourages portability of software with the use of common APIs, Drivers, and programming models.

On the telephony side of things, SCSA covers the discipline of access to algorithms and world-wide interfaces. This is enabled as a result of the standards in the SC message and on-board microprocessor control at the signal computing resource level. These signal computing elements such as DTMF detection, Call Progress, and world-wide line interface control, are all now able to converge in the same platform. This confluence is enabled because algorithm vendors and core technologists are mixing and matching their knowledge better than ever before with this standard. This allows advanced signaling methods and intelligent network features to be accessible to core technologists easily.

In call processing, SCSA enables the disciplines of specialty modular algorithms, such as those developed for speech recognition, speech synthesis, and compression technologies. The SCSA architecture encourages more levels of value-added in call processing algorithm development through common device drivers and APIs. In addition, the architecture begins to address system reliability issues for call processing platforms. These system reliability issues include the ability to respond deterministically to network alarms and requests, hot pluggability of nodes, and clock fallback capabilities. These disciplines were once the exclusive domain of large switch manufacturers, but now these elements of system reliability are available to SCSA designers.

Elements Required For Advanced Applications

This section is a high-level overview of mission-critical system support, host resident applications, and unified message and line interface processing. These a few of the elements that are sought after in building advanced applications. The application examples illustrate how these elements are leveraged in some way.

Mission-Critical Systems

The word "mission-critical" sounds imposing, but it's really a straightforward concept: *The System Can't Go Down.* There are many instances where customer requirements included "zero network downtime." Examples include military communications, high-volume transaction processing, and telephone company operations. All of these situations demand a level of service that is not tolerant of downtime. The applications is serving a life-or-death, service bureau, or multiple transactions of a revenue-producing nature.

SCSA offers a number of capabilities to designers of such systems. For example, voice messages can be redundantly stored on multiple signal computing nodes due to the timeslot broadcasting capability of the SCbus and the multi-node capability of the SCxbus. In addition, the SCxbus offers "hot pluggability" of nodes. That is, that a signal computing node, (which may have as many as 128 ports of voice processing capability) can be taken off-line and another node can be plugged into its place without affecting the other nodes or causing them to go off-line.

Of course, the software required to gracefully transfer or shut-down processes on the affected node before taking off line is critical here, and I'm not suggesting that the entire capability is a function of firmware. Nonetheless, the hot pluggability issue is very important in mission critical environments so that service can remain on-line while moves, changes, and upgrades are occurring.

Host Resident Applications

Many advanced applications require software to run in a "native" host environment. In other words, programmers in a development environment which is natively a (Tandem Guardian) environment are accustomed to the conventions and rules of that non-stop computer. It is rare that these individuals are going to re-train just to put a call processing application on line. This is why its important for applications to be developed in whatever environment suits the user or developer.

SCSA supports this with the Server API specification. The Server API allows Switch-Host-Server integration, so that developers can work in their native environment. Many advanced applications require central application control from either a mainframe or minicomputer. This is especially the case in high-volume transaction processing. Such is the case with cable pay-per-view, stock portfolio inquiries, and home banking, for example.

Call processing technology and voice processing in general provide a flexible front end to mainframes and workstations. This is why SCSA will bring a whole new generation of solutions to the communications market -- because SCSA makes call processing, telephony, and computing come together.

Unified Message and Line Interface Processing

The ability to unify messaging elements is also critical to a number of advanced applications. This is because there is such a dizzying array of message types and media which could possible be merged. There's e-mail, voice mail, fax mail, letters, telephone calls, beeper messages, and so on. It's very difficult (especially for travelers) to pick-up, review, and act on this mix of messages. Call processing systems are perhaps the industry's only hope in truly integrating these messages, and allowing users to freely mix and match both the sending and retrieval of same.

Users need both local & remote access to technologies that allow them to trade these same messages. For example, a business traveler may have a portable computer, as well as access to voice mail and fax mail. The challenge is to provide a unified "front end" to all of these messages so that the user can "self navigate" through all of these messages in one transaction.

SCSA provides a means to cost-effectively access these technology resources. For example, fax cards, voice cards, and data cards that provide associated message access capability can be housed in the same unit in high densities. This is cost effective, because it cuts down on the number of raw technology "ports" required to service a group of users. For example, if 20 voice, fax, and data ports were housed together for use by 40 people, it is certainly less expensive than buying separate technologies of the same type for that same number of people. This assumes, of course, that the constituents do not all require simultaneous access to the same resources.

SCSA Advanced Application Examples

The following applications have some relevance to the communications market because they combine computing, telephony, and call processing disciplines. In this section, I will cite examples in:

- Mission-critical messaging
- Choke-proof transaction processing
- Remote multi-media access
- Heterogeneous OLTP / common user interface
- Enterprise-wide call control

Mission-Critical Messaging

The challenge:

Building reliable, high-density systems with technologies that meet sophisticated user demands

SCSA addresses this challenge for a variety of users. For example, large companies that need to provide messaging capabilities messaging for many sites. In addition, telephone company-based messaging requires high density systems, as do service bureaus. Here are the success factors and associated SCSA offerings that make building mission critical messaging systems possible:

Success Factors:	SCSA Offers:
• Fault tolerant application control	• Server API
• Message redundancy	• Timeslot broadcasting
• Zero network downtime	• Clock fallback
• Hot pluggability for nodes	• SCxbus

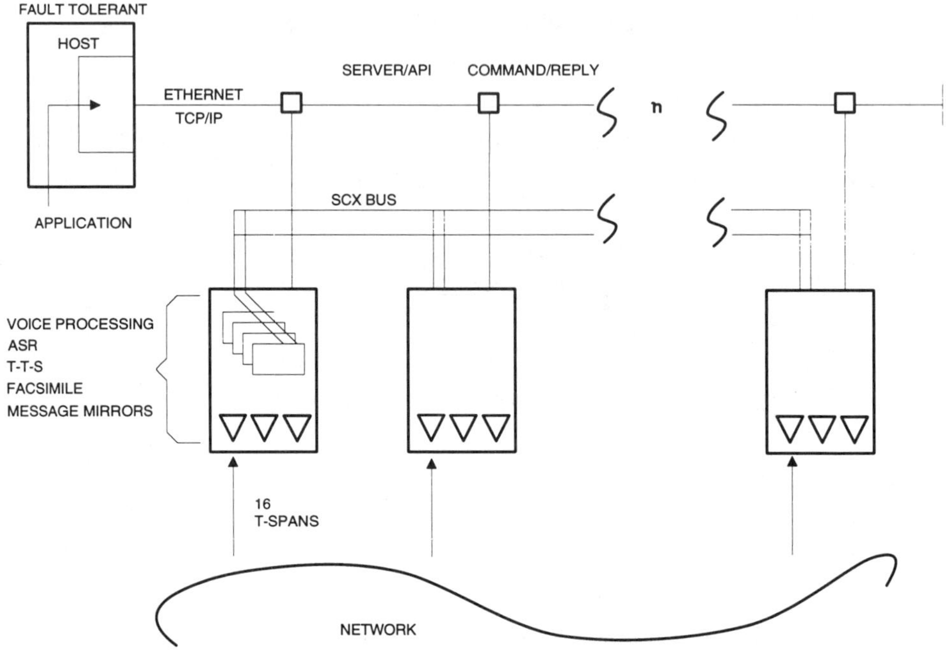

Figure **(MAGOO09.DRW)** illustrates how multiple signal computing nodes can be networked using the SCxbus. In addition, the Server API is used to control the array of nodes from a fault tolerant computer. Each node can house a variety of technologies including voice processing, ASR, Speech Synthesis, and Fax.

Some of the SCSA features such as clock fallback, timeslot broadcasting, and hot pluggability are useful in this kind of environment. Chapter 5 discusses the technical details of how these capabilities are implemented.

Choke-Proof Transaction Processing

The challenge:

> ***Automate traffic load balancing with call processing***

Not all users of services are able to use their own terminal or special device to access a company's database or information offering. Many times, the demand is so great for information or products, that telephone lines to the order desk are clogged for hours or days. Call processing systems can be used to "spread out" the traffic load for high-transaction environments. In this example, signal computing nodes are geographically distributed and located in key user cities.

By concentrating the nodes in the areas of the constituents' need, a "point of use" type of model can be deployed. This approach treats the call processing gear as remote peripheral equipment, while centralized processing of the application can occur elsewhere. This is especially critical if the same "remote peripheral equipment" is to be used by multiple applications.

The need for "choke proof" type systems is the most acute for companies who conduct nationwide polling, pay-per view & catalog sales, and centralized financial services. Here are the critical success factors and associated SCSA offerings that attack this challenge:

Success Factors:	**SCSA Offers:**
• Localize high density traffic at local nodes	• Out-of-box balancing with SCxbus
• Localize intra-node traffic with multi-technologies	• SCbus and SC message
• Aggregate high-volume traffic into smaller transactions	• Server API remote link

Figure **(MAGOO12.DRW)** illustrates what is meant by choke-proof transaction processing. On the left side of the diagram, geographically distributed signal computing nodes collect and service traffic from different areas of the country. These "slaves" are generic front-end processors to centralized applications.

The host processor can control the nodes via X.25 or other datacom means, while commanding the units to play out and record messages that are specific to a variety of applications. For example, the host processor may be running access programs to banks, cable companies, and department store order desks all at the same time.

FIGURE MAGOO12.DRW

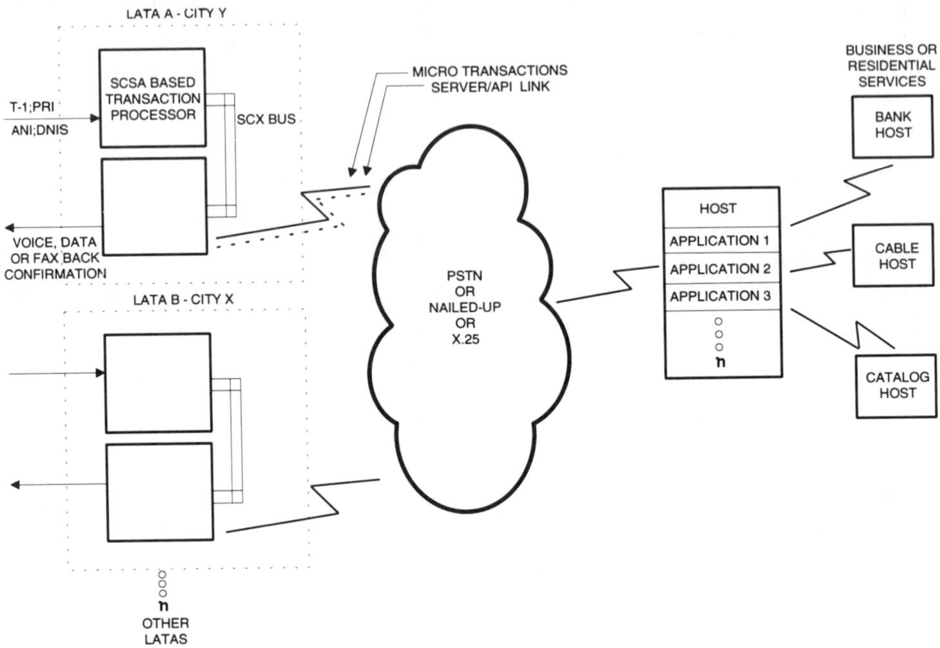

This arrangement can be extremely cost effective if a service bureau wants to achieve an economy of scale in deploying voice processing gear. This topology also cuts down on the number of circuits required to haul voice traffic. The voice traffic is "suppressed," or pushed back to the local node, so that only critical database access and ordering information has to flow back to the host processor.

Remote Multi-Media Access

The challenge:

Improve non-real time information exchange

There are a variety of proposed standards for video and data transmission over wide areas. These include Px64, JPEG, MPEG, and others. And there are some impressive video compression and video workstation technology coming out of Compression Labs, PictureTel, and Total Multi-Media, to name a few.

What's missing, however, is an entire infrastructure to manipulate the switching, storage, and access to those transmission paths. Also absent is a scheme for the true networking of video along with other call processing elements. While SCSA is not proposed as a video standard per se, there are a number of SCSA elements that make both real-time and non-real-time communication mere effective.

SCSA addresses this challenge for casual use by knowledge workers in so-called "work groups," traveling PC users, and by companies who want to provide total information access to their customers. Detailed below are some of the critical success factors for remote enterprise Multi-Media access, and some of the associated SCSA offerings that will address these concerns.

Success Factors:	SCSA Offers:
• Link workstations & portables with access to many technologies	• SCbus to bind technologies
• Integrate messaging applications between E-Mail, FAX-Mail, and Voice-Mail	• SCdpi & SCapi abstracts application from specific devices
• Economize on centralized functions	• Server API enables client/server paradigm

Figure **(MAGOO15B.DRW)** illustrates how a work station can be equipped with a variety of multi-media technologies. In this example, the workstation not only has video, voice, and fax processing capabilities, but also a station interface and switching card, so the first workstation can be a local call controller for a number of telephone extensions.

The SCxbus can be used to link multiple workstations together. In essence, these work stations are signal computing nodes with specialized client software. The link between the nodes may be some other new standard, such as FDDI II or ATM, for example. I'm posing SCxbus as a viable alternative because of the way it enables the active sharing of a variety of technologies on a high-speed bus. It may be the case several years from now that the physical, electronic, and data link layers on SCSA are consistent with those required to support ATM and FDDI II for SCxbus multi-node implementations.

FIGURE MAGOO15B.DRW

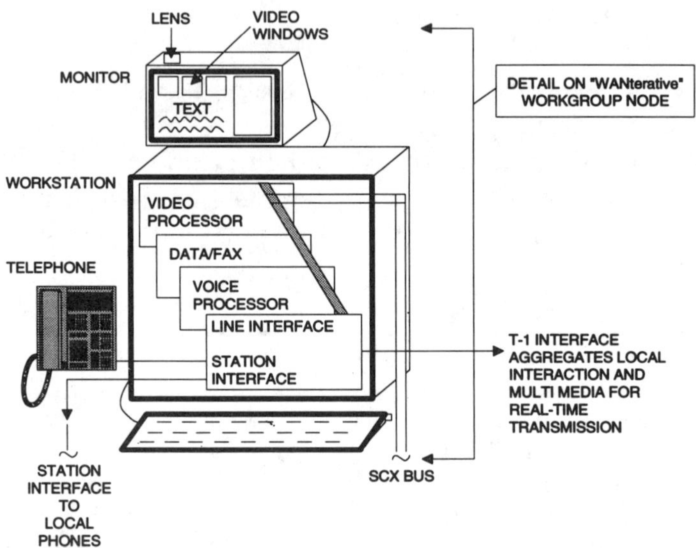

Figure **(MAGOO15C.DRW)** illustrates how multiple "WANteractive Work groups" can be networked together over digital facilities. This enables real-time video processing sessions, telephone conferences, and fax broadcasting between sites, for example.

Heterogeneous OLTP / Common User Interface

The challenge:

> ***Provide universal access to services with disparate terminal devices and transport methods***

It's difficult to accommodate the number of access methods and terminals that users are accustomed to. Today, even more choices of terminals are coming down the pike. For example, ADSI (screen-based) phones, personal digital assistants, portable PCs, digital cellular telephones, portable faxes, and character-based terminals of all types.

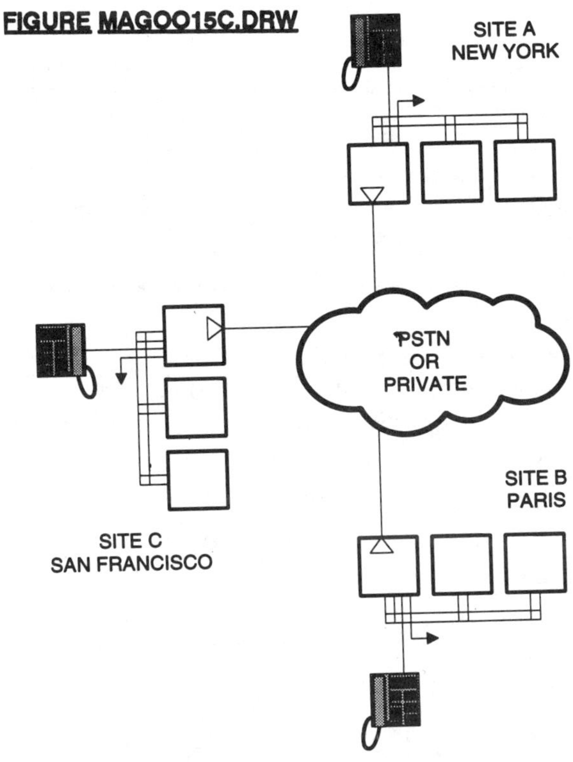

Those who provide services for home banking, flight status and scheduling, and stock portfolio applications want to broaden their service reach as much as possible in order to appeal to the widest audience. SCSA enables this because many technologies are being brought together in the same platform.

For example, Dialogic has cards that provide both simple voice store and forward capability, and also ADSI protocol. The same company, along with Brooktrout, and GammaLink, are providing combined voice and fax functions. There are already data compression, and terminal emulation boards available -- tie to these the line interfaces to the network and the SCbus -- and you've got a means to provide universal access to multiple terminals.

The table below describes the critical success factors and associated SCSA offerings that will help to make this new class of OLTP (on-line transaction processing) system available:

Success Factors:	SCSA Offers:
• Provide universal terminal access for ADSI, Voice, FAX, 327x	• SC message for flexible use of algorithms
• Consolidate network traffic of unlike terminals	• SCbus and transmission over digital facilities
• Flexible front-end to mainframes & minis	• Server API

Figure **(MAGOO18A.DRW)** illustrates how OLTP-based services can provide access to clients with varied terminals. The SCSA-based OLTP controller handles the data handling and protocol conversion for a variety of communications and then communicates this to the host processors. These hosts can control the OLTP controller as a peripheral I/O device, or application intelligence can reside in the controller, in which case the host and controller are peers.

FIGURE MAGOO18A.DRW

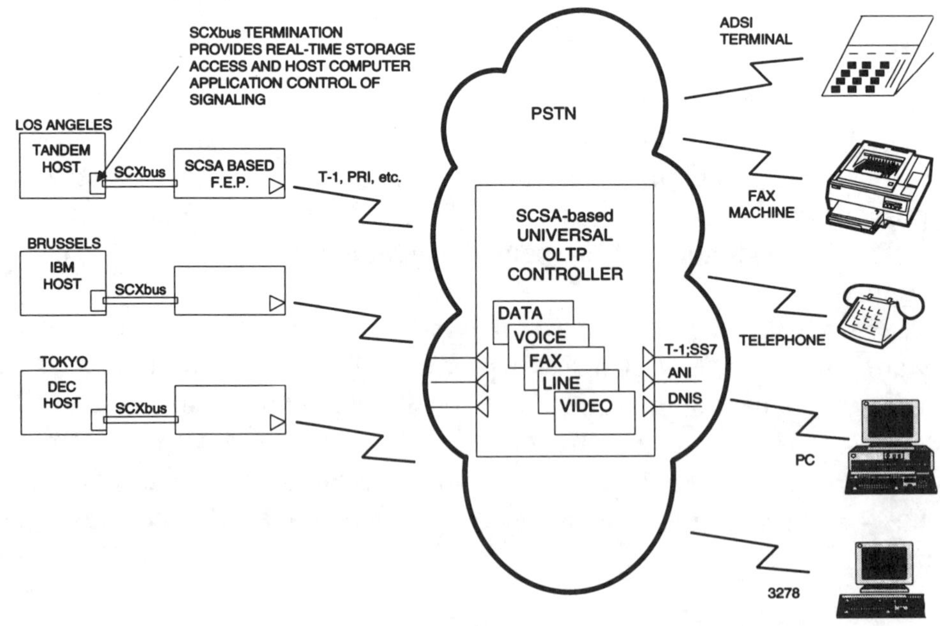

In this scenario, a T-1 drop and insert configuration is being used in the SCSA-based system. This allows real-time transactions to be aggregated in the OLTP controller, treated for screening, and then passed to the host processor for further action only if required. SCSA-based Front End Processors can be co-located with each host and linked via the SCxbus, so that real-time retrieval and storage of critical messages can be handled by the host itself.

Enterprise-Wide Call Controller

The challenge:

> ***Achieve "GLOBAL ACD" capability for large enterprises while protecting investment in PBXs, ACDs and private networks***

Multi-site companies with many different switches have a set of unique challenges on their hands. They are often service-based business with peak demands, and they may require emergency routing and traffic off loading from time to time. This is difficult to achieve due to the fact that so many telephone switches are utterly incompatible with one another.

Long distance carriers such as AT&T, Sprint, and MCI have tried to risc up to this challenge by offering value-added network services such as coordinated dialing plans, on-net and off-net access routing, and electronic switched tandem networking of geographically distributed switches.

What with the advances in DSP technology, call processing, and communications programs available in the voice response area, the question is often begged: "What do we need these old switches for anyway?" Although some of us would like to scrap the old way of doing business in communication, and simply replace it all with SCSA-based communications controllers (that do it all), the reality of imbedded equipment, and the associated investment in yesterday's switching technology is a constant.

An interim solution is the Enterprise-Wide Call Controller. Some of the success factors and associated SCSA offerings are detailed in the table below:

Success Factors:	**SCSA Offers:**
• Mission-critical control of agent transfer	• Server API remote link
• Interoperability	• SCxbus adjunct links between switches
• Provide access to many combined technologies	• SCbus & SCapi

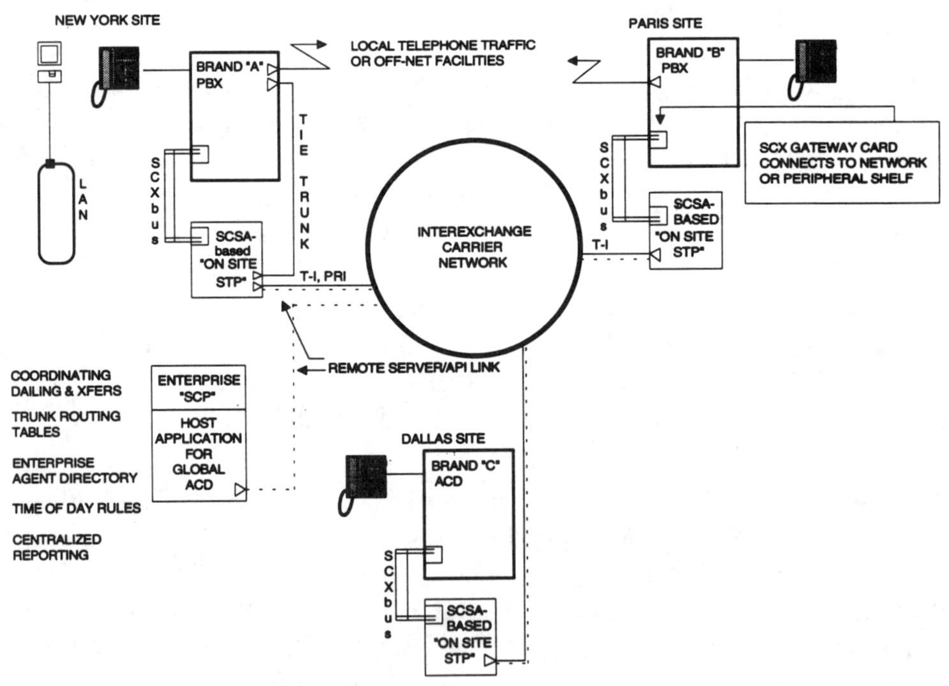

Figure **(MAGOO21.DRW)** illustrates how SCSA-based components can be used as a private SS7 (Signaling System Seven) - like network. For example, on-site SCSA-based "Signal Transfer Points" can hook-up to a PBX back plane via the SCxbus. This is entirely doable, since virtually all PBX manufacturers are using some kind of PCM/TDM back plane that would at least be electronically compatible with the SC 2000 chip. In fact, the Siemens PCM Highway, and the MITEL STbus are indeed pin-compatible and electronically compatible with the SC 2000 chip. Mitel's' STbus is the bus of choice for no less than a dozen switch manufacturers, so this is not as far off as one would assume on the surface.

Should there be a lack of cooperation on the part of a certain switch manufacturer to provide the hooks to do this, then a direct T-1 or PRI interconnection could be achieved, so that the SCSA-based "STP" could act as an access tandem or PBX on a set of "tie" trunks. This concept was discussed in the first chapter.

The SCSA-based "STP" would also be hooked-up to a centralized host computer, which would run the routing, dialing plan, agent rules, and directory information for the enterprise. This host, in effect, would be acting as the Signal Control Point (SCP) for the transactions going in and out of the enterprise. Control of the on-site STPs can be achieved via a Server API link over X.25 or switched 56K facilities.

A typical call scenario would involve a person calling a site on a regular outside (local) telephone line. The trunk routing tables inside the PBX would be set-up to automatically route the call to a TIE trunk (which is tied to the SCSA-based STP), or directly to a back plane connection to the SCxbus of the STP. The STP would then listen for any call control, identifying digit, or voice commands from the person calling into the PBX.

If the person speaks a name, or dials digits of a certain extension number, then the call would be routed to that extension. In this scenario, the same call could be switched to another site over digital facilities. This would be done under the control of the SCP, since it would have a state table, which kept track of what "agent" was available at what site. In this fashion, the switching, routing, and application control of the communications for this enterprise are in the control of the enterprise itself, not a PBX switch generic.

This configuration allows users and customers to make one telephone call, which can then be attached and re-attached to whomever, or whatever the caller wants to communicate with, rather than having to hang-up and make repeated and separate calls. A transaction that required several voice communications, a fax pick-up, and voice messaging could be achieved in this way.

Summary

Above all, teamwork is required to solve current and future communication problems. SCSA provides architects of communications systems with a platform that encourages this teamwork, but its people that will make unified systems a reality, not an architecture itself. As you have seen in this chapter, SCSA leverages the best of 3 worlds:

- Unified application control of disparate devices and vehicles
- Flexible telephone network access and control
- Access to top call processing core technologies

There are many rewards that will come out of working with SCSA endorsers. Some of the new and existing communications systems architects will see a rush of opportunities for new products and services, have a standard platform for global solutions, and be able to source innovative software applications from many vendors in order to bring their offerings to market quicker.

The next chapter explains how this is all possible from a technical standpoint.

CHAPTER 5 - SCSA IN DEPTH: THE TECHNOLOGY

This chapter provides an in depth overview of the SCSA technical specifications. This chapter is a collection of selected text and functional descriptions for the modules that define the Signal Computing System Architecture. Each module description is in a particular phase of development. These phases describe the level of detail in the document, and define the process through which each description must pass before being proposed as a standard. The phases are:

Development Phase	Document Content	Input/Feedback Sources
CONCEPT	Proposed functional requirements.	Selected SCSA users.
WORKING	Proposed functional descriptions, and in some cases, interface specifications.	A Working Group of interested parties.
EVALUATION	Proposed functional descriptions, interface specifications, and implementation guidelines.	Registered users of this SCSA Technical Specification Document (contact Dialogic for a current copy)

Each section of this chapter is currently in the phase described below.

Chapter Section	Phase
1. SCbus Specification	Evaluation
2. SCxbus Specification	Evaluation
3. SCmessage Functional Description	Working
4. SCdpi Functional Description	Working
5. SCapi Functional Description	Concept
6. Server API Functional Description	Concept
7. System Integration Description	Working

Overview

As the demand for multiple technologies and sophistication in multi-port telephony-based systems has increased, the need for a cohesive platform development environment has become more critical. To this end, Dialogic and other industry leaders have announced an architectural infrastructure for the development of extensible multi-port telephony-based systems. This architectural infrastructure is called Signal Computing System Architecture.

The phrase platform development environment encompasses a number of concepts and activities including the sourcing of core technologies for multi-port systems including *voice store and forward, facsimile, text-to-speech, and voice recognition.* In addition, the platform development environment describes the way in which a vendor, or group of vendors cooperate with customers to assist in the development of systems across architectures, operating systems, and specific applications of a technology.

The goal of a sound platform development environment is to ensure the maximum return on investment to an enterprise's efforts in developing successful products. The overall durability of investment needs to be protected not only in hardware that is purchased, but also in application-specific software. Each development effort, then, must take a forward view of how a design can adapt to future customer needs in terms of system functions, sizing, interoperability with other systems, and coupling with newly introduced technology.

The Signal Computing System Architecture addresses this goal by providing an industry-wide structure for multi-port systems development which spans technologies, operating systems, and platform sizing. Signal Computing System Architecture provides a unified approach to the overall design of multi-port telephony-based systems for the voice processing and call processing industry. The common use of integrated device drivers, complimentary technologies, and applications code between several architectural implementations will lower development cost, speed time to market, and provide a stable point of reference for this growing industry.

By adopting the Signal Computing System Architecture model, systems developers can reduce the duplication of effort in application and software tool development. In addition, developers can avoid dead-end architectures that do not provide for integration of technologies when needed, or modular growth for several ports to hundreds and even thousands of ports.

Existing applications may evolve into Signal Computing System Architecture compliance in phases. Applications that comply with *SCSA* may do so at any tier of the OSI-like model which is presented here. Third-Party development of hardware and firmware elements may comply with *SCSA* at any tier as well, however, the extensibility and interoperability between platforms and architectures are what allows third party elements to comply more completely with Signal Computing System Architecture.

Improving Systems Development of Signal Computing Platforms

Signal Computing System Architecture provides a focal point in the voice and call processing industry through which Dialogic, other hardware vendors, development tool providers, and customers can cooperate in order to advance the consistency of an industry-wide platform development environment.

As core technologies such as *voice store and forward, facsimile, text-to-speech, and voice recognition* become increasingly available from numerous technology-based companies, the cost of keeping pace with advances in technology has increased. The primary challenge presented to application developers and systems integrators is to plan for the growth and evolution of systems design in such a way that it can adapt to a highly volatile market which demands higher sophistication every day.

SCSA supports industry-wide standardization towards a consistent platform development environment. This development consistency will ensure both speedy access to new and improved core technologies and access to market opportunities through the extension of software development efforts. Signal Computing System Architecture supports three main areas for achieving consistency in an industry-wide platform development environment:

- Integration of Technologies on multi-port Signal Computing Systems
- Extensibility of Software Development across Architectures
- The Tenets of Open Systems Design

Integration of Technologies on Multi-Port Telephony-Based Systems

An exciting array of technologies is now available which allow for the conversion of text messages into the spoken word, the speaker verification over the telephone, and integrated facsimile and voice transactions, to name a few. The approach in combining these technologies in such a way that they are accessible during the same transaction is difficult, and doing it in an economical fashion is even more of a challenge.

There are several approaches to this challenge, including the logical arrangement of multiple technologies to provide for a 1:1 ratio of use, and the ultimate combining of technologies either discretely on the same board, for example, or through firmware inter-coupling of the technologies. All of these approaches are valid, however, some implementations are insensitive to the *incremental growth needs* of a platform's technologies.

Signal Computing System Architecture provides for a way to:

- Add technologies as needed, with minimum investment for basics

- Design exclusive or shared use of technologies on a port-by-port basis

- Migrate the use of core technologies across platform architectures

Signal Computing System Architecture defines the most flexible approach for the use of multiple technologies on a multi-port telephony-based platform, so that developers developers can create systems that are adaptable to the coupling of technologies on an as-needed basis.

Extensibility of Software Development across Architectures

Today, those involved in the development of multi-port telephony-based systems must use a variety of firmware interfaces, hardware drivers, application development tools, and communications protocols. To be effective in overall systems design, and especially to extend the software investment into future iterations of the platform, developers must learn the peculiarities of each hardware element and its associated driver interface. Developers are often forced to make software adaptations to ensure that the communication between the application software and hardware is enabled regardless of the disparate nature of hardware drivers. This problem is compounded even further when an application is required to run on a variety of platforms which use separate architectures.

Signal Computing System Architecture defines a way to:

- Develop applications based on an integrated device driver and API scheme
- Migrate the bulk of application code from one architecture to another
- Decouple Applications to run independently of the signal computing node

By developing multi-port telephony-based systems based on Signal Computing System Architecture, developers can design application-specific software and tools which can work across multiple platform architectures.

The Tenets of Open Systems Design

Modern customers demand flexibility in their application design. Most applications solve particular business problems, so developers don't want to create new problems, such as *systems inflexibility*, along the way. Today, multi-port telephony-based systems use a wide array of software, firmware, and hardware from Dialogic and other industry participants. It is Dialogic's position to support Open Systems Design and this is central to its business philosophy. This includes the idea of *interoperability* between platforms, *extensibility* of architecture design, the *scalability* of features and functions within an architecture, and the general *portability* of software across architectures. These same concepts were first identified and held-up as industry-wide icons by organizations such as UNIX International (UI) and the Open Systems Foundation (OSF). These same concepts owe much of their hopeful implementation to a migration to standards-based design.

Dialogic's *OPEN Development Program (OPEN)* is an example of an industry focal point for the support of the Signal Computing System Architecture. With over 70 software and hardware participants, *OPEN* acts as a clearing house for third-party hardware development, application development tools, and application software providers. The program provides for technology information transfer, joint development, and interactive forums to accelerate solutions to the marketplace for multi-port telephony-based systems.

Signal Computing System Architecture propels the advance of an industry-wide platform development environment by embracing a variety of standards and solutions from the world of telephony, computation, and data communications. Signal Computing System Architecture will empower an ever-increasing body of developers by providing guidelines for systems development based on accessible and easy to use architectures.

Dialogic will make advances in its own products and services which comply with the chief tenets and structure of the Signal Computing System Architecture. In addition, Dialogic will continue to publish and make available both documentation and tools which will encourage a variety of vendors to develop extensions and additional capabilities to Signal Computing System Architecture.

The Open Systems View

The open structure of the SCSA hardware interfaces, firmware interfaces, and driver environments will encourage industry-wide consistency that can deliver new technologies both today and in the future. SCSA Specification follows the OSI model and provides:

- Documented interfaces between layers of the architecture
- Description of hardware, firmware, software elements
- Revision control by Dialogic
- Nominal documentation fee for updates

Figure **(OSIVIEW.DRW)** displays a model which approximates how SCSA conforms to the OSI model. This is one way of looking at the architecture which provides a common ground for developers on the commutation side and developers on the telephony side to establish a "common language" for working together in SCSA environments.

The model displays the client computer, signal computing nodes, and the telephone network. In addition, all of the important interfaces between these elements are displayed. This chapter will go into detail on the interfaces in this model. The most current revision of the hardware, software, and firmware specifications for SCSA is available from Dialogic, however, this chapter should go a long way to getting you familiar with how it all works. The chapter will cover the most significant input and output points on the OSI view of the SCSA architecture, starting at the lowest (bus) level, and working up to the API level.

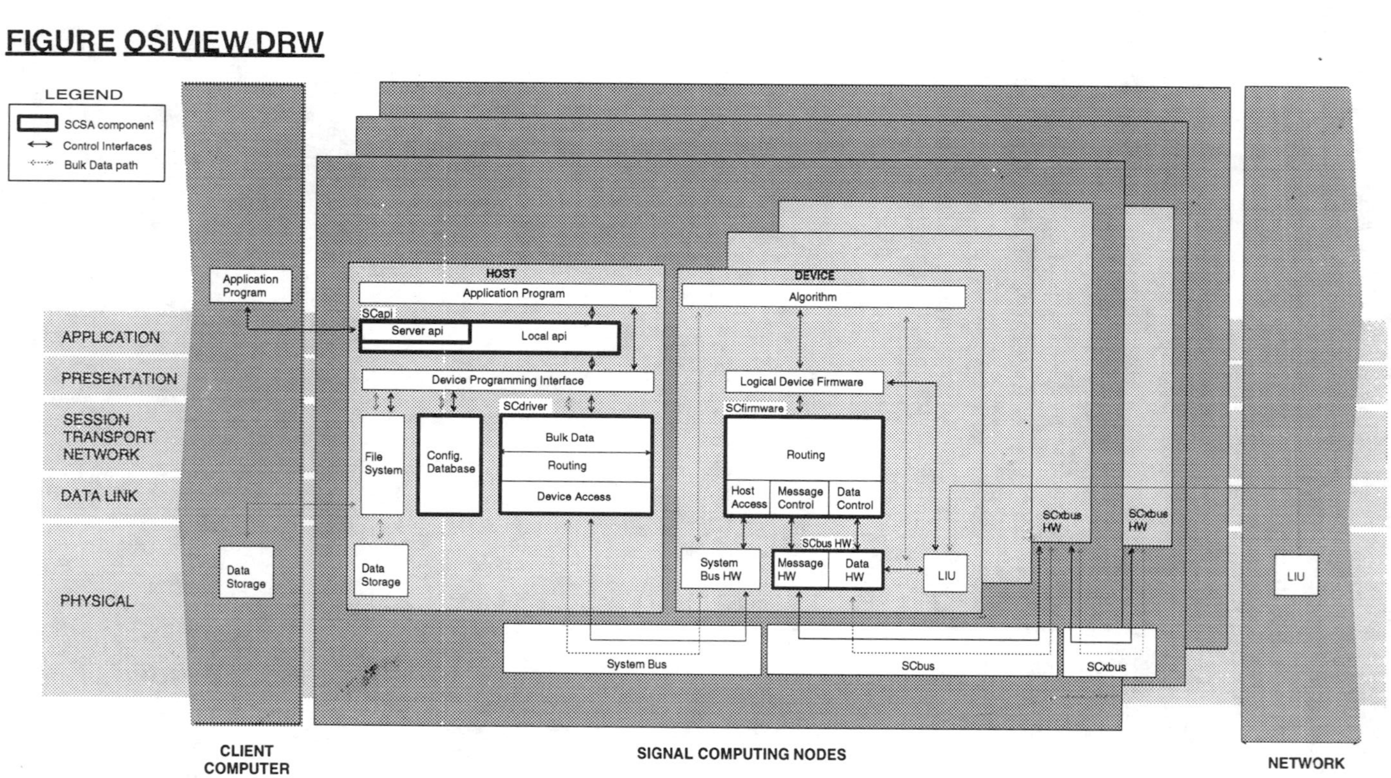
FIGURE OSIVIEW.DRW
LEGEND
SCSA component
Control Interfaces
Bulk Data path
APPLICATION
PRESENTATION
SESSION
TRANSPORT
NETWORK
DATA LINK
PHYSICAL
Application Program
Data Storage
HOST
Application Program
SCapi
Server api
Local api
Device Programming Interface
SCdriver
File System
Config. Database
Bulk Data
Routing
Device Access
Data Storage
DEVICE
Algorithm
Logical Device Firmware
SCfirmware
Routing
Host Access
Message Control
Data Control
SCbus HW
System Bus HW
Message HW
Data HW
LIU
SCxbus HW
SCxbus HW
System Bus
SCbus
SCxbus
LIU
CLIENT COMPUTER
SIGNAL COMPUTING NODES
NETWORK

SCbus and SCxbus Key Features Overview

Figure **(SCBUS.DRW)** is a blow-up of the SCSA architecture snapshot which concentrates on the SCbus itself. This is the bus where all data, telephone, video, fax, and other traffic flows inside the call processing platform. It connects all of the Call processing resources together over multiple data lines.

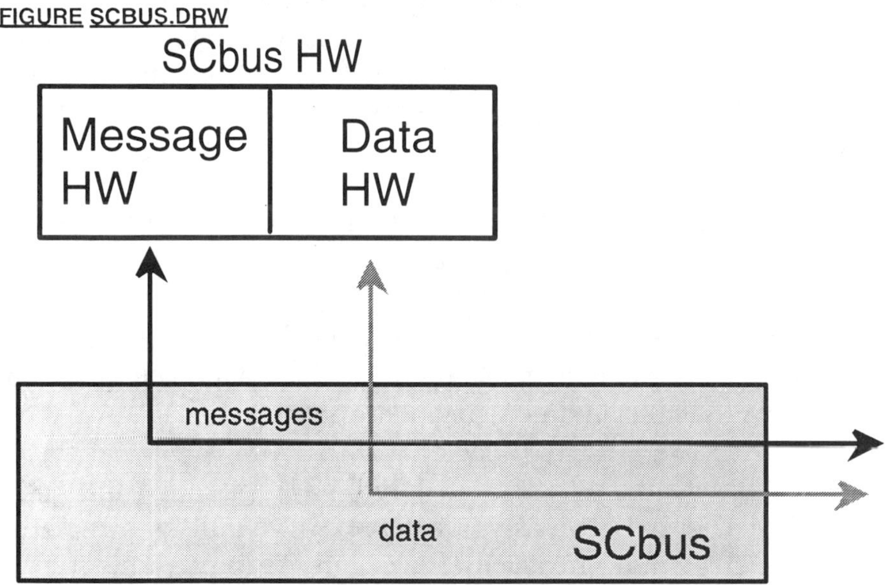

The SCbus represents a design breakthrough over previous generation buses with its use of out-of-band signaling protocol and ability to carry information at higher speeds. A signal processing system typically requires four classes of data transfer; board-to-board high speed bulk data transfer, board-to-board message transfer, CPU-to-memory high speed bulk data transfer, and CPU-to-board message transfer. The bus architecture is a hybrid in that it is made up of a synchronous TDM bus, referred to as the data bus, and a serial message passing bus, referred to as the message bus. The SCxbus is a means to extend these busses to accommodate multiple signal computing nodes.

In terms of some concrete benefits for embracing the SCbus in call processing system design, here are a few points that developers are excited about:

Increased bus speed provides more time slots for higher density systems means more time slots for collateral technologies and higher densities.

Most customers who wish to create high-density systems are in dire need of a higher speed bus, so that they are not forced into multiple platform topologies, or forced to go with some short-lived proprietary solution. The SCbus solves that problem. The bus is worthy of use not only in PCs, but also in tandem with large-scale computing platforms that can process thousands of transactions per second - such as OLTP, or on-line transaction processing systems. In addition, developers need to have access to multiple technologies per call.

The bus itself handles 16 data lines, which are supervised via a common signaling channel. This is to a certain extent, an in-box method of implementing an *SS7* (signaling system seven) type of model, where various transport channels are controlled by a data channel which acts as a traffic cop. This is in our opinion, a much more modern approach to building sophisticated and efficient call processing platforms.

The standard bus rate for the SCbus is 4.096 MHz, although the bus can handle 2 and 8 MHz speeds. So as you can see, we've built-in some growth to this bus in anticipation of greater user needs in the future. This translates into *512, 1024 or 2048 universal time slots (1024 standard).* What's immediately apparent to developers about the capabilities of the SCbus is how it makes a quantum leap in terms of its ability to handle many more timeslots than any currently available bus in our industry.

The number of time slots supported in standard mode is 1024. What that means is that over 1000 high-speed communications can be occurring at the same time on the bus. Now, if you want the conversations or communications to be real time between two people or full duplex, for example, that equates to 512 full duplex channels. And none of these channels are wasted on signaling or "unused" as with past generation busses.

The SCbus allows developers to mix and match technologies in any ratio they desire. For example, a four-port fax card, a 16-port voice recognition card, and a two-port Text-To-Speech card can be shared across let's say 4 T-Spans. Or, a one-to-one ratio can be applied to the same system if this is required.

Simple and powerful timeslot control including non-blocked switching and broadcast between all time slots.

This time slot control allows also for an optimal switching and broadcast mode capability where you can broadcast one time slot over all the rest of them for conferencing or other functions. This is a superior implementation over previous generations, because broadcasts are available over the entire spectrum of 1024 timeslots.

When it comes down to just what you can plug-in from a standard telephony standpoint, the SCbus allows as many as 16 T-1 spans or 16 E-1 spans to be carried all in the same box. An E-1 span carries 30 discrete communications. So if you take 16 times 30, you can see how the math works out to a very robust communication highway.

The effective throughput of the bus in its maximum mode is up to 130 Mbps synchronous data throughput (64 Mbps Standard). Now that's the kind of throughput necessary to carry some serious traffic for high-density, mission-critical applications. It is now possible, because of this design to build large call center systems which mix voice, data, video, and other traffic for enhanced network services that many of our customers have been banking their futures on.

Electrically compatible with PEB, Mitel ST Bus and Siemens PCM Highway.

The bus is also electrically compatible with PEB, Mitel STbus (including MVIP) and the Siemens PCM highway. So, in terms of *speeds and feeds*, that's really superior to any open standard that's in use for efficient call processing.

Full frame buffering and time slot synchronization, so time slots can be bundled for clear channel data transport.

One of the most distinctive advantages of building platforms based on the SCbus is this timeslot bundling capability. Remember that with a 64 kbps timeslot, that's bandwidth of a regular telephone line; that's about the bandwidth required for a voice transmission or a fax transmission. But what about non-compacted full-motion, full color, real-time video? What about Bulk Data Transfer? The idea here is that you have the ability to bundle these individual timeslots together on the fly, so that your system can accommodate any kind of high-speed transmission that is required.

This enables a whole new age of microprocessor-based call processing. That's the whole idea with the architecture. It doesn't matter what you're sending across the pipe. As long as your technology conforms to the protocols and takes advantage of the Supervisory channel, you can set up and tear down clear channel and bundled communications at will. Now that's something that is simply not addressed by any of the current first and second generation busses.

Out-of-box expansion capability for multi-node systems and connection to other computers and switches.

Developers have struggled for a standard way to achieve large system densities using standard components and application development tools. Signal Computing System Architecture is positioned as an industry focal point for platform development in this area.

Currently, systems developers and integrators have taken a dual approach to the multi-node challenge. This includes the use of front-end switching and back-end LAN technology to provide a single view of multiple nodes or multi-port telephony-based systems. Another approach has been to link each node to a central computer, where an application resides to control the array as one device.

In addition to the high bandwidth supplied by the SCbus inside of one platform, the bus is designed also as an inter-node bus between multiple co-located call processing nodes. This allows for the design of multi-node systems that have a single-system view. The SCxbus specification of SCSA is positioned to standardize on methods for achieving this end while protecting as much of the original hardware and software investment possible. This will allow systems integrators and developers to provide solutions across the broadest range of system sizing in the industry today.

In effect, developers will be able to build systems that can accommodate thousands of lines, and allow resource sharing in-between the nodes for special feature access. The capability to build system redundancy is also enabled because of this bus extension capability.

For example, let's say you build a node and fill it with call processing resources and perhaps 10 T-spans coming into it from the network. Let's say you have a second system with the same number of network interfaces. Here's a way that SCSA will enable these two nodes to talk together - for a real-time transaction to go into the first one and travel out the back and scoot down the expansion bus and back into the second node. This way, a second telephone transaction to access the second node from the network is not needed.

Automatic clock fallback handles network glitches without service disruption

One of the most severe drawbacks of previous generation buses, is that clock fallback capability was either totally absent, or not very flexible from a programming standpoint. With the SCbus, a bad clock on a primary resource card can be automatically switched over to a secondary clock source. This allows for a higher degree of up-time for mission critical applications.

Detailed Technical Descriptions of the SC busses

The SC data bus which is supported by the SC 2000 ASIC, is the mechanism for inter-board bulk data transfer. The sixteen synchronous TDM data lines of the data bus provide a guaranteed bandwidth for such transfers. The data bus is specified to run at 2.048 Mbps, 4.096 Mbps, or 8.192 Mbps per data line. The smallest switchable unit on the data bus is the time slot. A time slot consists of 8 consecutive bits of data. Time slot data is transmitted in frames.

The number of time slots in each frame depends on the choice of data bus bandwidth, and is either 32, 64, or 128 time slots per frame, however the bandwidth equivalent of one time slot will remain fixed at 64 Kbps. A data path with a bandwidth of 64 Kbps is referred to as a "Digital Service level 0", or a DS0.

Figure **(SCBUS05.DRW)** displays how in its default 4.096 Mbps speed, the bus itself conforms to standard PCM/TDM (Pulse Code Modulation / Time Division Multiplexing) conventions. In the figure, the 16 data lines are represented in 16 "space domain" lines, while the "time domain" is fixed at 64 Kbps per data line, each having 64 timeslots. The 16 data lines are serviced by the SC 2000 ASIC switching block.

FIGURE SCBUS05.DRW

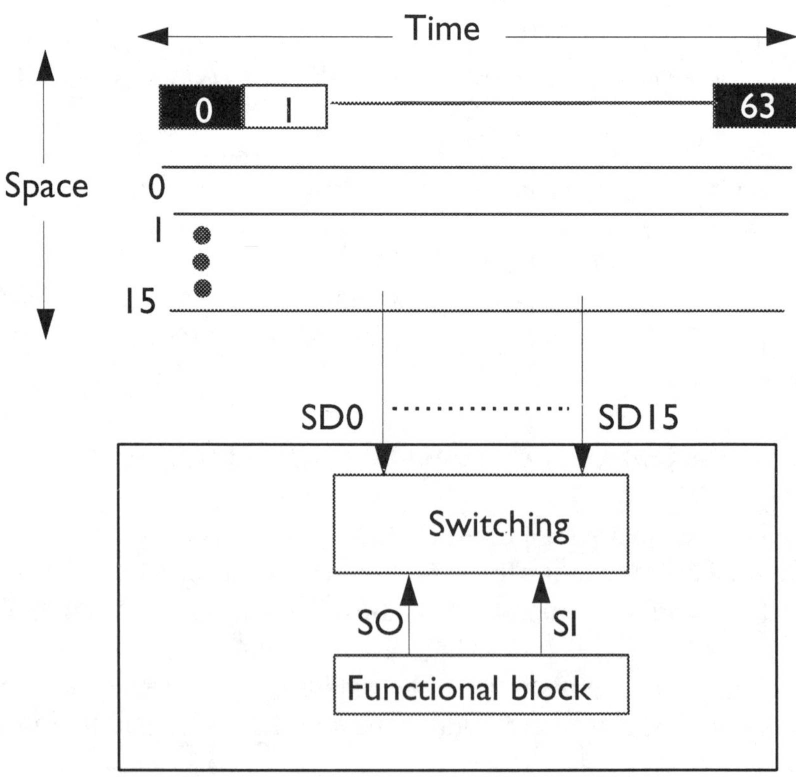

The SC message bus, which is supported by the HSCX (HDLC) chip, is the mechanism for inter-board message communication. It may also be used as a simple, low cost alternative to an extended, parallel system bus. The message bus is a multi-master (contention) bus, specified to run at 2.048 Mbps. The table below shows how the SCbus compares with previous generation resource sharing buses.

Capability	SCbus	MVIP	PEB
Bandwidth - number of time slots	2048*	512	96
Time slot utilization - input or output	Programmable	Fixed	Fixed
Number of T-1/E-1 trunks per box	32*	8	4
Separate common signaling channel	Yes	No (one time slot per channel)	No (inband robbed bit)
Time slot bundling (grouping multiple DS0s)	Supported	Requires additional circuitry	Not available
Loopback diagnostics on the bus	Supported	Requires additional circuitry	Not available
Automatic clock fall back	Supported	Requires additional circuitry	Difficult to implement
System expansion - multinode system	Via internode SCxbus extension	Not specified	Not specified
*8.192 MHz bandwidth			

SC Bus - System Architecture and Bus Structures

Figure **(SCBUS02.DRW)** shows the relationship between SCSA resource modules and the SCbus. In addition, the SCxbus adaptor is pictured as a module which extends the Data Bus and Message bus to other signal computing nodes.

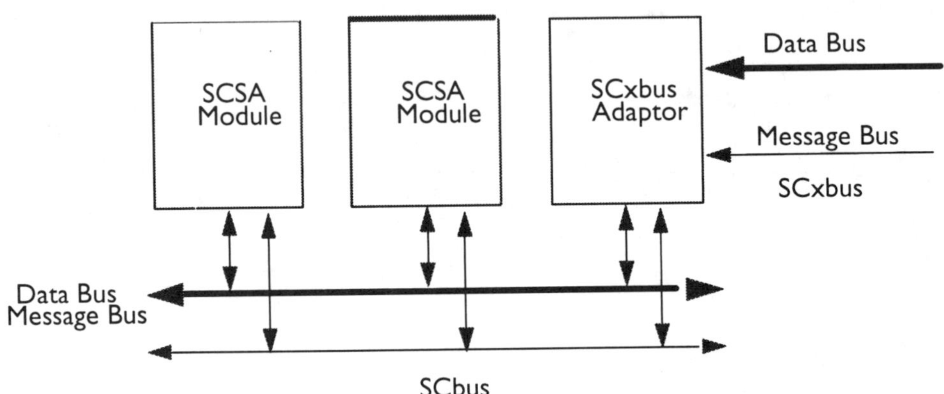

SCbus Data Path Description

The SCbus data path is a point-to-point link between any two ports without signal loss. This data path has the following characteristics:

- Only one port may transmit data during a time slot period.
- Any number of ports may receive data during a time slot period.
- There is no restriction on the values of the data that can be transferred over the data bus.
- Time slots may be bundled to create data paths with multi DS0 bandwidths.
- Any time slot may be used for transmitting or receiving data.

The routing software running on the host CPU is responsible for maintaining the relationship between logical devices. The routing firmware running on the device CPU is responsible for executing resource management commands issued by the host routing software, and for maintaining the relationship between time slots on the SCbus and time slots on the board's internal TDM bus.

FIGURE SCBUS04.DRW

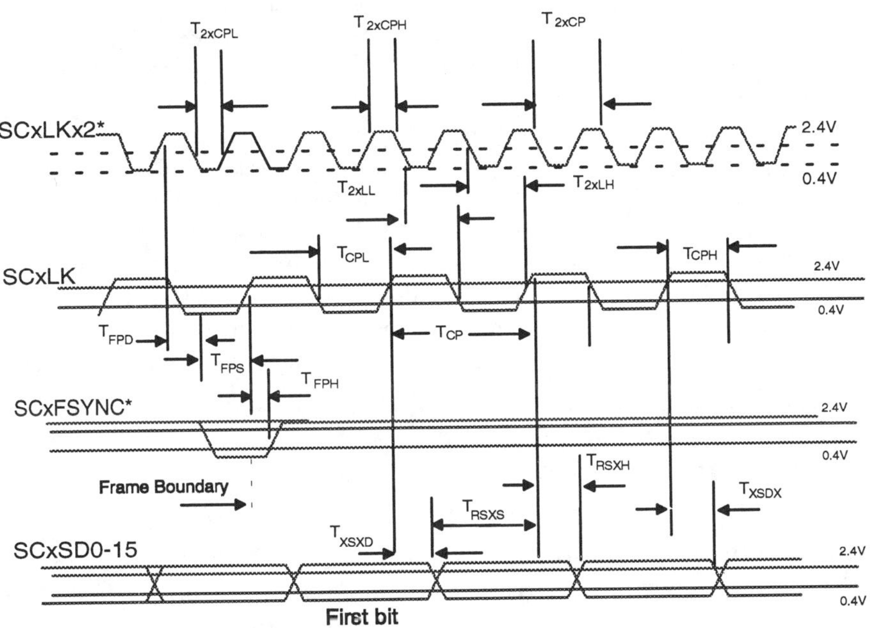

Figure **(SCBUS04.DRW)** shows the Data Bus timing diagram for the SCbus.

Symbol	Description	Explanation
SCLK	System Clock	This signal is driven by the clock master, and is used to read serial data from, and write serial data to, the data bus. Data transitions occur on the positive going edge of this signal. The frequency of SCLK is specified to be either 2.048 MHz, 4.096 MHz, or 8.192 MHz. The frequency of SCLK may be selectable.
SCLKx2*	System Clock-by-Two	This signal is driven by the clock master and is synchronous to, and twice the frequency of, SCLK. Transitions of SCLK occur on the falling edge of SCLKx2*.

FSYNC*	Frame Sync	This active low signal is driven by the clock master, and is used to indicate the start of a data frame. FSYNC* goes low for one SCLKx2*-period, one half bit-period prior to the first bit of the first time slot of the frame. The period of FSYNC* is 125ms.
SD[0..15]	Serial Data	These sixteen lines are driven by individual boards on a time slot by time slot basis. Any one board may drive any line for any time slot period. Each of the lines, SD0 to SD15, may also be referred to as a stream.
CLKFAIL	Clock Fail	This active high signal is driven by the clock master, and must be driven with open collector drivers. During normal operation of the system clocks CLKFAIL is held low. If the clock master detects the failure of, or the loss of Synchronization within, the clock source it releases CLKFAIL, which is then pulled high. CLKFAIL is used to initiate the reassignment of the role of clock-master from one board to another.

Message Bus Interface

The message bus is a multi-master serial bus running at the fixed rate of 2.048 Mhz. This section discusses the physical interface to the bus. The SCbus specification includes an out-of-band signaling capability using HDLC protocol. This allows developers to take advantage of a recognized standard in signaling and apply it with off-the-shelf components, like the Siemens HSCX (HDLC) chip.

Message Bus

MC Supervisory Signal. This line is a serial data line and must be driven with open collector drivers. Data is shifted onto this line with a 2.048 Mbps clock, referred to in this reference as MC clock. MC clock IS NOT available from the SCbus, and must be generated on every board that intends to use MC.

The MC clock

An MC clock signal is generated by every board which is capable of accessing the message bus. MC clock is defined to be a 50% duty cycle signal running at 2.048 MHz, synchronous to SCLKx2*. Data is shifted on to the MC line on the rising edge of MC clock, and read off the MC line on the falling edge of MC clock. The relationship between MC clock, FSYNC*, and SCLKx2* is shown below. Note that in this example SCLK is running at 4.096 MHz. A divide-by-two circuit is needed to derive MC clock, which must run at 2.048 MHz.

FIGURE SCBUS_9.DRW

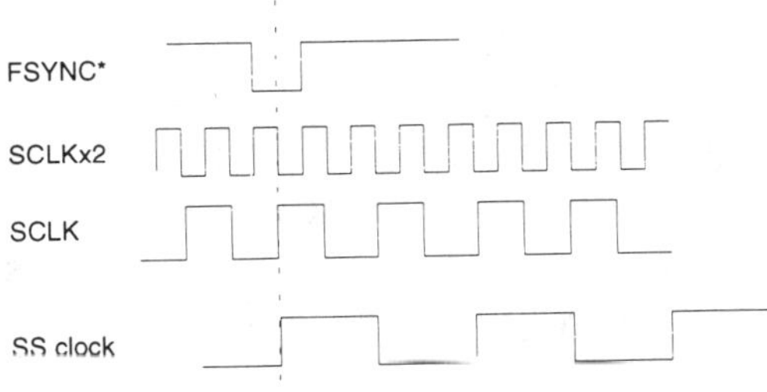

The Bus Idle State

When a board has no data to transmit over the message bus it must present a high impedance interface to the MC line. When the bus is pulled high for more than 8 consecutive bit periods (i.e. when more than 8 consecutive 1s are detected on the message bus) the bus is said to be in the IDLE state. In order to prevent the transmission of false IDLE patterns the MC interface will transparently insert 0s ("bit stuffing") in accordance with the specifications of the HDLC protocol.

Message Frame Format

Messages sent over the message bus are arranged in frames. Each message frame is defined by a pair of identical flags, which begin and terminate the frame. This frame flag is specified to be the binary sequence (01111110) (hexadecimal 7E), in accordance with the HDLC protocol. A message bus message frame must not be confused with a data bus data frame, which is defined by FSYNC* alone.

Message Frame Format

The minimum frame length is 9 octets, the maximum frame length is specified to be 128 octets. The message frame contents are yet to be completely specified but will most likely conform with the CCITT HDLC protocol.

Contention Resolution

Each board that is capable of transmitting data on the message bus must be able to monitor the state of the bus, simultaneous to such a transmission. Bus contention is detected when a board shifts a 1 ("high") on to the bus, but simultaneously reads a 0 ("low"). Bus contention is resolved with the following procedure:

1. The first board to detect contention must immediately abort it's message frame, and set it's message bus interface to the high impedance state.

2. The board will not attempt retransmission of the aborted frame until it sees the message bus is in the IDLE state.

3. Once a board has transmitted a full message frame it must not attempt a further transmission until it has detected 10 successive 1s on the message bus.

FIGURE SCBUCONT.DRW

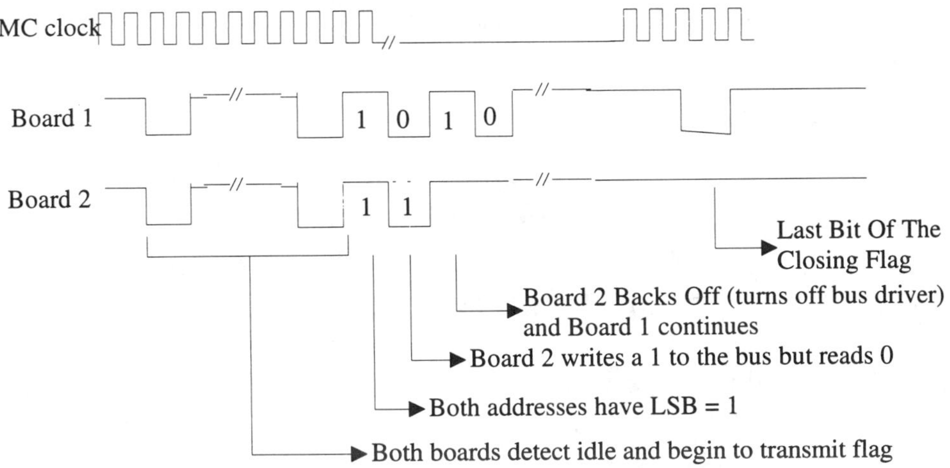

Board ID

Each SCSA board is required to have a unique ID within the system. This ID number must be configurable, for example by use of a switch, and must be accessible to the device firmware

Data Bus Interface

The data bus is a space-time division multiplexed serial bus running at 2.048, 4.096 or 8.192 MHz, allowing point to point connections between a transmitting board and one or more receiving boards to be established. This section discusses the physical interface to the data bus. For information on Initialization see the "System Integration" discussion at the end of this chapter.

Time Slot Uniqueness

Only one board is permitted to drive the data bus during any particular time slot period.

Universal Time Slots

Any time slot may be used to transmit or receive data.

Multi-time Slot Capability (Hyperchannel)

The data bus hardware interface may be capable of transmitting or receiving data distributed over multiple time slots, that is the data path may be composed of multiple DS0s.

Bit Integrity

Digit or bit sequence integrity should be maintained for the length of a DS0 transmission path. Bit integrity is considered to be maintained when the binary values in a byte transmitted by a board are exactly reproduced at the receiver. Connections using robbed-bit signaling and controlled slip are exceptions. The maintenance of bit integrity is the equivalent of a 0 db transmission loss, therefore the data bus should appear as a 4-wire transmission path with no signal loss within the system.

Byte Integrity

Byte sequence integrity should be maintained for the length of the DS0 transmission path. Byte integrity is maintained when the sequence of bytes in a frame remains the same between source to destination. Controlled slip is an exception. Byte integrity is equivalent to time slot integrity.

Multi-time Slot Integrity

Time slot sequence integrity should be maintained for a multiple time slot (multi-DS0) path. Time slot sequence is considered to be maintained when a set of time slots within a frame retain the same relative positions within the frame, from source to destination.

Clear Channel Capability

Data paths are to be bit-sequence independent. There should be no restrictions on the number of consecutive 1s and 0s, or any binary pattern, able to be transmitted on a DS0.

Timing Signal Synchronization

Because the data bus does not use acknowledgments, or any other form of handshaking, to synchronize data transfers it is important that the relationships between timing signals and data are strictly adhered to. There are three important timing relationships: bit, time slot, and frame. Bit Synchronization establishes the proper timing for sending or receiving data. Time slot and fame Synchronization ensures that operation of all data bus interfaces are synchronized. The following rules must be followed:

1. Data must be shifted on to the bus on the rising edge of SCLK.

2. Data must be shifted off the bus on, or after, the falling edge of SCLK.

3. The time slot counter must be reset during the FSYNC* pulse.

4. The time slot counter must be incremented or reset on the rising edge of SCLK.

Data Frame Format

Each serial data stream is divided into frames. Each frame is further divided into 8-bit subframes called time slots. The first time slot is designated time slot 0 (TS0), the second TS1, etc.

FIGURE SCBUS_3.DRW

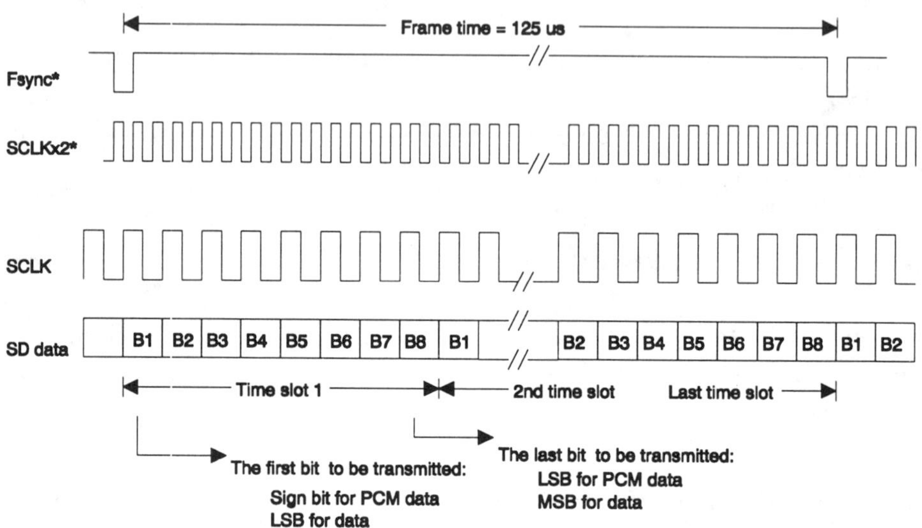

The first bit in each used time slot is designated B1 and is transmitted or received first. The last bit in a time slot is designated B8, and is transmitted or received last. This bit numbering scheme adheres to CCITT conventions. For a PCM coded byte B1 is the sign bit, and B2 the MSB (Most Significant Bit) of the byte. For binary data bytes B1 is the LSB (Least Significant Bit).

Unused Time Slots

If a board is not writing data to the data bus it must present a high impedance interface to the bus. In this way time slots which are not being used will be pulled high.

Energy Levels

The coded speech energy levels on the data bus must comply with FCC regulations. Therefore it is required that the average coded speech energy level transmitted on an SCbus channel may not exceed -12 dbm0 when averaged over a 3 second period. The average coded network signaling energy level transmitted on an SCbus channel may not exceed -3 dbm0.

Data Transfer

The data bus is a synchronous, TDM, serial bus. Data transfers over the bus are accomplished by assigning one or more time slot IDs (for example a data bus stream number, plus a time slot number) to the transmitting and receiving boards. During the assigned time slot period(s) the transmitting board will drive the bus and the receiving board will read data off the bus.

A transmission path that transports data between two end points is referred to as a channel. The interface point between a channel and a board is known as a port. When a channel connects two ports a call exists. In the SCSA standard the device which implements channels, and performs the mapping between SCbus time slots and a board's internal TDM bus time slots is called the SCbus Hardware.

The data bus architecture supports three classes of data transfer. These are discussed in the following sections.

Symmetrical Data Transfer

Symmetrical data transfers occur between two ports when they both send data to, and receive data from, the other.

FIGURE SCBUS_4.DRW

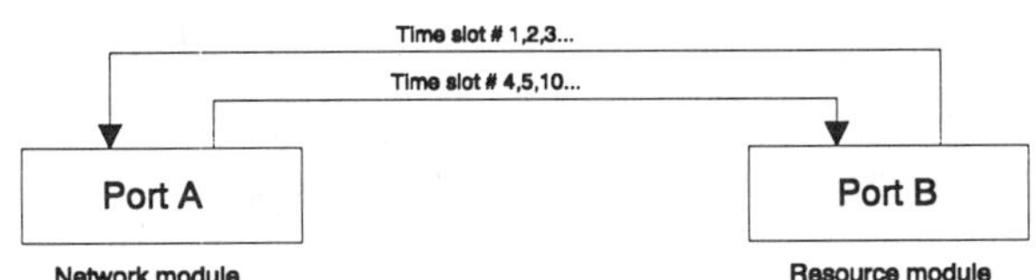

Asymmetrical Data Transfer

Asymmetrical data transfers occur when a port receives data from a port other than the one(s) it is transmitting to.

FIGURE SCBUS_5.DRW

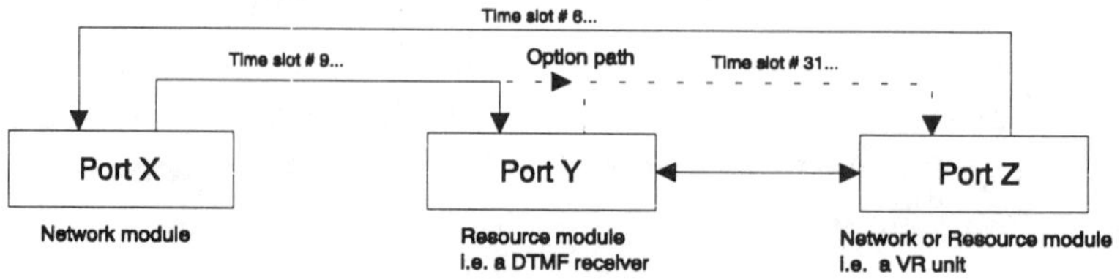

In this example port X uses time slot 9 to send data to port Y, and time slot 6 to receive data from port Z.

Broadcast Data Transfer

Broadcast data transfers occur when one port transmits, and multiple ports receive, on one channel; that is when a port establishes a transmission call to several ports simultaneously.

FIGURE SCBUS_6.DRW

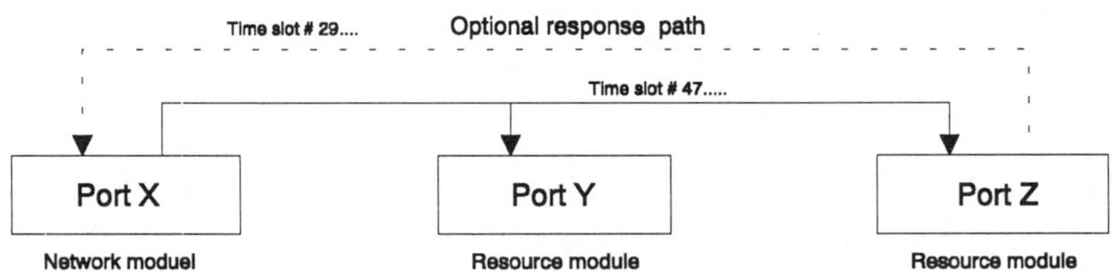

In this example port X broadcasts data on time slot 47 to port Y and Z. Z may respond using time slot 29.

SCbus Timing Specifications

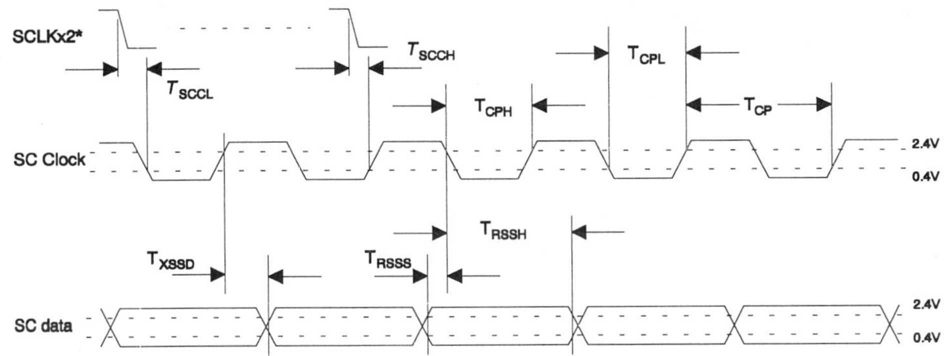

Symbol	Description	Min	Typ.	Max	Unit
T sccl	SCLKx2* low to MC clock low delay			25	ns
T scch	SCLKx2* low to MC clock high delay			25	ns
T cph	Clock period high at 2.048 MHz	237	244	252	ns
T cpl	Clock period low at 2.048 MHz	237	244	252	ns
T cp	Clock period at 2.048 MHz	481	488	496	ns
T xssd	Output data delay - from clock going high to valid output on the bus.			55	ns
T rsss	Input data set up time -- from valid data to clock going low.This means that MC data is valid and stable at least 124 ns before the clock falling edge.	124			ns
T rssh	Input data hold time - from clock going low until data change. This means that MC data will remain valid at least 224 ns after the clock falling edge.	124			ns

Data Bus

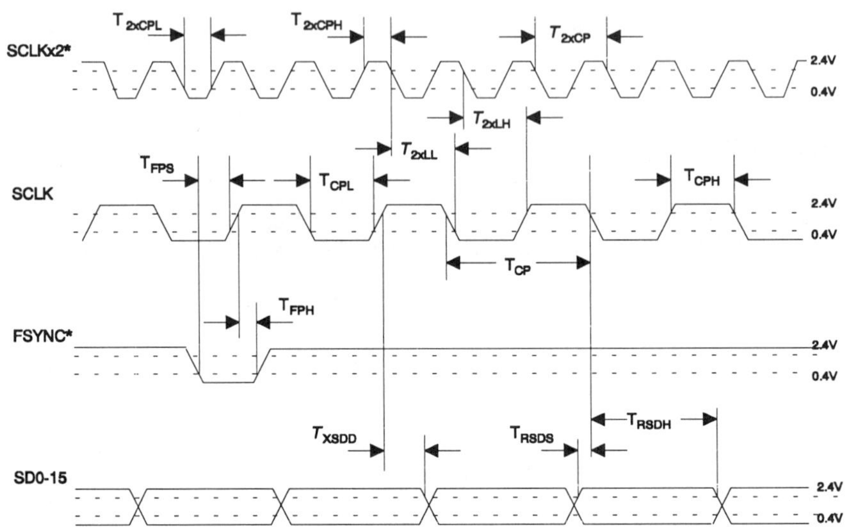

DATA BUS TIMING TABLE SCLK = 2.048 MHz

Symbol	Descriptions	Min	Typ.	Max	Unit
T 2xcph	Clock period high at 4.096 MHz	115	122	130	ns
T 2xcpl	Clock period low at 4.096 MHz	115	122	130	ns
T 2xcp	Clock period at 4.096 MHz	237	244	252	ns
T cph	Clock period high at 2.048 MHz	237	244	252	ns
T cpl	Clock period low at 2.048 MHz	237	244	252	ns
T cp	Clock period at 2.048 MHz	481	488	496	ns
T x2ll	4 MHz clock low to 2 MHZ clock low delay.			20	ns
T x2lh	4 MHz clock low to 2 MHz clock high delay.			20	ns
T fps	Frame pulse delay - from SCLKx2* clock going high to valid output on the bus.			60	ns
T fph	Frame sync pulse hold	119			ns
T xsdd	Output data delay - from SCLK going high to valid output on the bus.			55	ns
T rsds	Input data set up time - from valid data to SCLK going low.This means that SD data is valid and stable at least 124ns before the SCLK falling edge.	120			ns
T rsdh	Input data hold time - from SCLK going low until data change. This means that SD data will remain valid at least 224ns after the clock falling edge.	T cpl + 25			ns

DATA BUS TIMING TABLE SCLK = 4.096 MHz

Symbol	Descriptions	Min	Typical	Max	Unit
T 2xcph	Clock period high at 8.192 MHz	54	61	69	ns
T 2xcpl	Clock period low at 8.192 MHz	54	61	69	ns
T 2xcp	Clock period at 8.192 MHz	115	122	130	ns
T cph	Clock period high at 4.096 MHz	115	122	130	ns
T cpl	Clock period low at 4.096 MHz	115	122	130	ns
T cp	Clock period at 4.096 MHz	237	244	252	ns
T x2ll	8 MHz clock low to 4 MHz clock low delay			20	ns
T x2lh	8 MHz clock low to 4 MHz clock high delay			20	ns
T fps	Frame pulse delay - from SCLK clock going high to valid output on the bus.			35	ns
T fph	Frame pulse hold	59			ns
T xsdd	Output data delay - from SCLK going high to valid output on the bus.			55	ns
T rsds	Input data set up time - from valid data to SCLK going low. This means that SD data is valid and stable at least 124ns before the SCLK falling edge.	60			ns
T rsdd	Input data hold time - from SCLK going low until data change. This means that SD data will remain valid at least 224ns after the clock falling edge.	T cpl + 25			ns

SCbus Electrical Specifications

General Specifications

The following overall restrictions should apply:

1. The physical length of the SCbus should not exceed 21 inches overall.

2. The total number of boards connected to the SCbus should not exceed 20.

Message Bus

All message bus drivers should be TTL open-collector type capable of sourcing 24mA minimum. It is recommended that Schmidt-trigger receivers be used on all timing signal inputs. The total number of message bus drivers should not exceed 40.

Parameter	Symbol	Min	Max	Unit
Operating Voltage	V_{CC}	4.75	5.25	
Input voltage - High	V_{iH}	2.00	5.25	V
Input voltage - Low	V_{iL}	- 0.25	.80	V
Input leakage current	I_{LH} or I_{LL}		+/- 10	μ A
Output High current at 2.4 V	I0H	- 0.40	-24	mA
Output Low current at 0.4 V	I0L	24		mA

Termination

The MC line should be pulled up on each board using a 4.7 KΩ 1/4 W resistor.

Data Bus

All data bus drivers should be TTL tri-state type capable of sourcing 24 mA minimum. It is recommended that Schmidt-trigger receivers be used on all timing signal inputs. The maximum load on any SCbus line should not exceed 150 mA and/or 15 pF (including connector) per board.

Parameter	**Symbol**	**Min**	**Max**	**Unit**
Operating Voltage	V_{CC}	4.75	5.25	V
Input voltage - High	V_{iH}	2.00	5.25	V
Input voltage - Low	V_{iL}	-0.25	.80	V
Input leakage current	I_{LH} or I_{LL}		+/- 10	μ A
Output High current at 2.4 V	I_{0H}	-0.40	-24	mA
Output Low current at 0.4 V	I_{0L}	2		mA
Output -off leakage current	I_{LZ}		10	μ A

Termination

Each stream (SD0 to SD15) should be pulled up by a 100 KΩ 1/8 W resistor. Only one pull-up resistor per stream should be used on any one board assembly, which may consist of a main board and one or more daughter boards.

Each timing signal (SCLK, SCLKx2* and FSYNC*) should be pulled up by a 100 KW 1/8W resistor. Only one pull-up resistor per signal should be used on any one board assembly, which may consist of a main board and one or more daughter boards.

CLKFAIL should be pulled up by a 4.7KΩ 1/4 W resistor on each board. A resistor of 47 or 51Ω may be placed in series with each output to improve the signal quality.

SCbus Mechanical Specifications

The mechanical specifications current at this time are relevant to ribbon cable (mezzanine) implementations of the SCbus only, and in particular implementations intended for the interconnection of PC/AT or PC/2 standard cards. The SC 2000 ASIC and HSCX chip set do not limit the SCbus to this particular implementation in other circumstances, where a different mechanical implementation (e.g. a back plane bus like VME) may be more suitable.

Cable and Connector

The SCbus uses straight 26 conductor "ribbon" cable as the physical connection between boards. The connector is specified to be a 26-pin right angle connector, Amphenol type 842-821-2699-895, or equivalent. Two polarization tabs are required per connector, Amphenol type 842-800-1611-003, or similar. The maximum length of the SCbus cable must not exceed 21 inches. The minimum distance between connectors must not be less than 1 inch.

FIGURE SCBUS_16.DRW

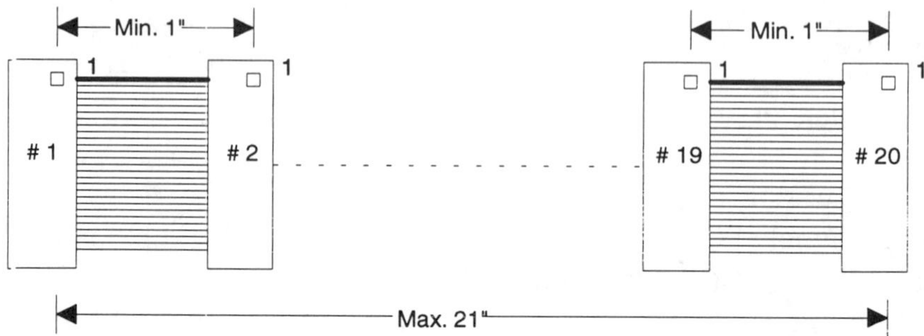

SCbus Connector Pin Out

Pin #	SCbus Sig.	Description
1	SCLKx2*	Twice the frequency of SCLK
2	GND	Connected to PCB
3	SCLK	System clock. Driven by clock master.
4	Reserved	
5	FSYNC*	Frame sync pulse. Driven by the clock master.
6	CLKFAIL	Clock fail. Active high. Driven by clock master
7	SD0	Bidirectional serial data stream # 1.
8	GND	Connected to PCB
9	SD1	Bidirectional serial data stream # 2.
10	SD2	Bidirectional serial data stream # 3.
11	SD3	Bidirectional serial data stream # 4.
12	SD4	Bidirectional serial data stream # 5.
13	SD5	Bi-directional serial data stream # 6.
14	SD6	Bi-directional serial data stream # 7.
15	GND	Connected to PCB
16	SD7	Bi-directional serial data stream # 8.
17	SD8	Bidirectional serial data stream # 9.
18	SD9	Bidirectional serial data stream # 10.
19	SD10	Bidirectional serial data stream # 11.
20	SD11	Bidirectional serial data stream # 12.
21	GND	Connected to PCB
22	SD12	Bidirectional serial data stream # 13.
23	SD13	Bi-directional serial data stream # 14.
24	SD14	Bi-directional serial data stream # 15.
25	SD15	Bi-directional serial data stream # 16.
26	MC Data	Message bus

Proposed connector positions on MicroChannel and ISA boards.

FIGURE SUBUS_17.DRW

Connector location for PC/AT (ISA) Bus

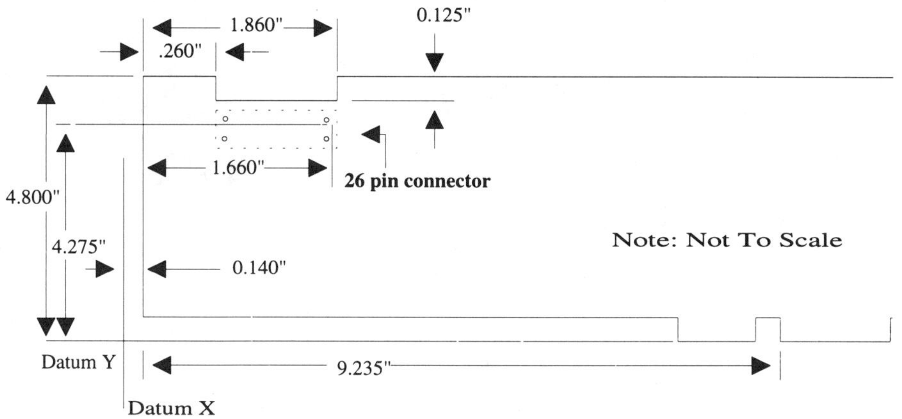

Connector location for MicroChannel Bus Type 3 Adapter

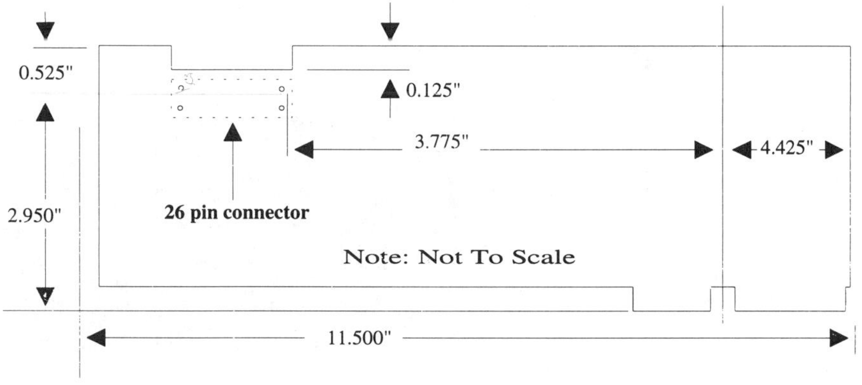

SCxbus Specification

SCXBUS High-level Diagram

FIGURE SCXBUS.DRW

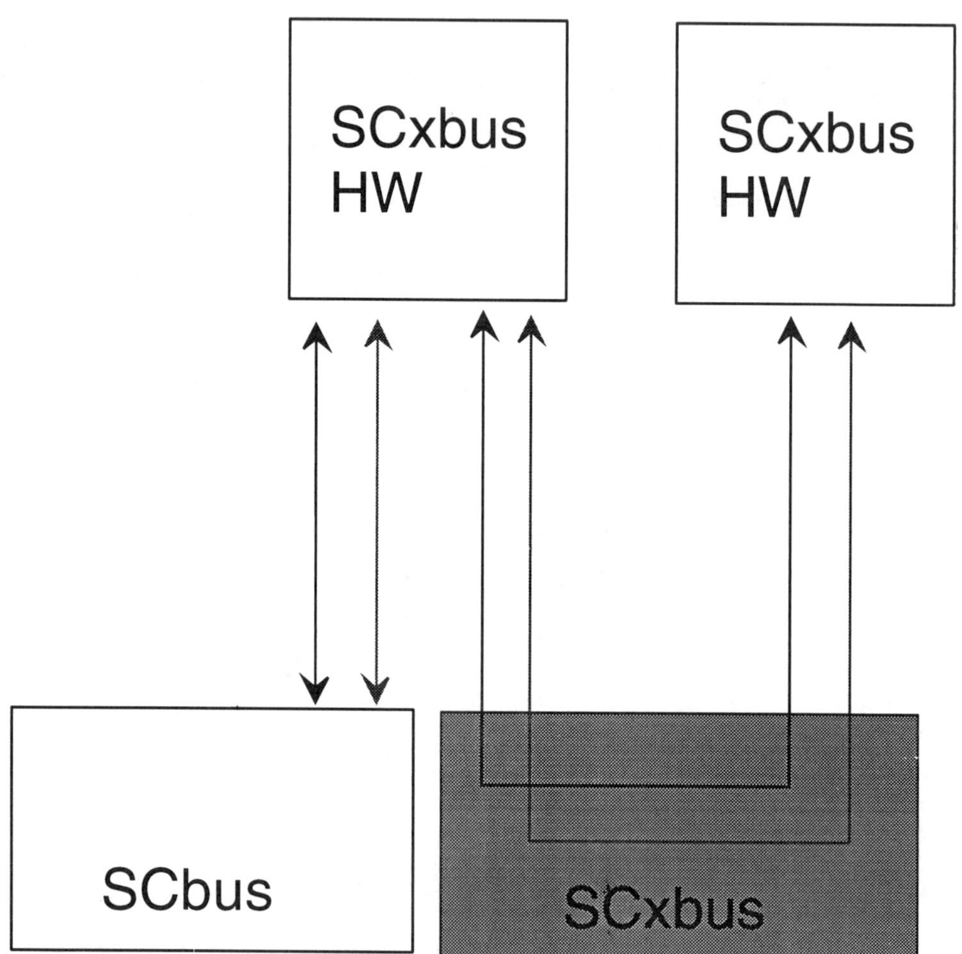

SCxbus Architecture Overview

The SCxbus (SCSA extension bus) is the inter-node equivalent of the SCbus. Up to 16 nodes ("boxes") can be connected to the SCxbus. The following figure illustrates a possible configuration. Note that each box has its own CPU.

SCxbus inter-node connection Diagram

SCBUS_18.DRW

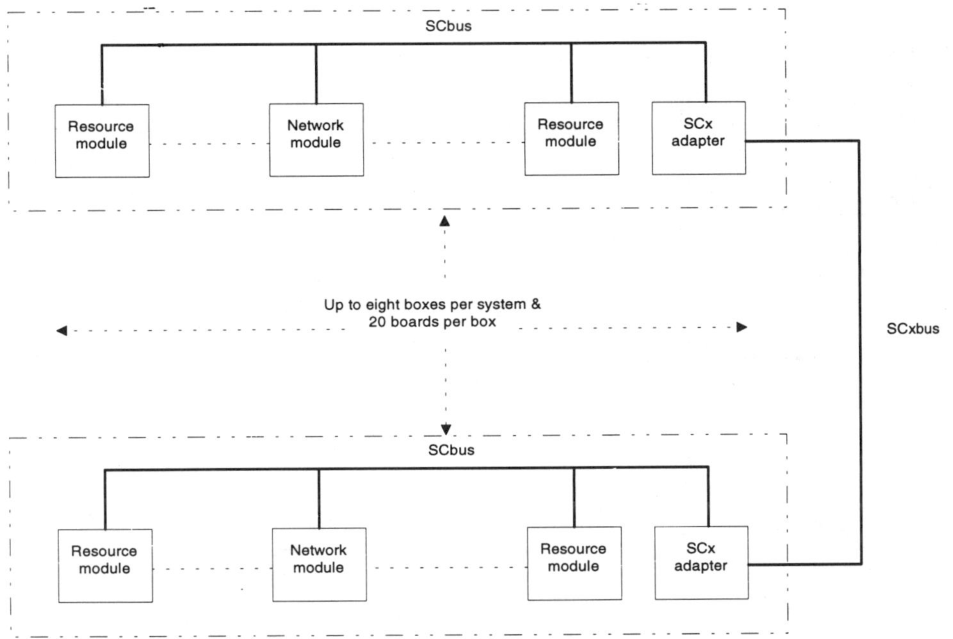

SCxbus Key Features Overview

SCxbus is an expansion bus which is used to connect multiple signal computing nodes. An SCxbus adaptor in each node is used to connect the nodes, which provides a single system image for high-density implementations.

The SCxbus specification calls for a 1024x1024 switching capability, thus allowing the same level of timeslot switching and resource sharing as is available on the SCbus. The expansion bus uses the same protocol as SCbus and has a fixed 4.096 Mbps bus speed with 1024 universal time slots.

An RS485 differential transceiver is specified for the bus, so its end-to-end cable length is limited to 50 feet. This means that its capacity of 16 nodes must be realized in rack-mounted or co-located chassis.

A node may be fitted with more than one SCxbus adapter to increase system throughput and reliability. Each node can be optionally connected to a LAN, in which case the LAN host can control the system remotely, viewing the system as a single unit. Figure (SCBUS_33.DRW) shows what such a system topology would look like.

SCXBUS LAN CONFIGURATION.

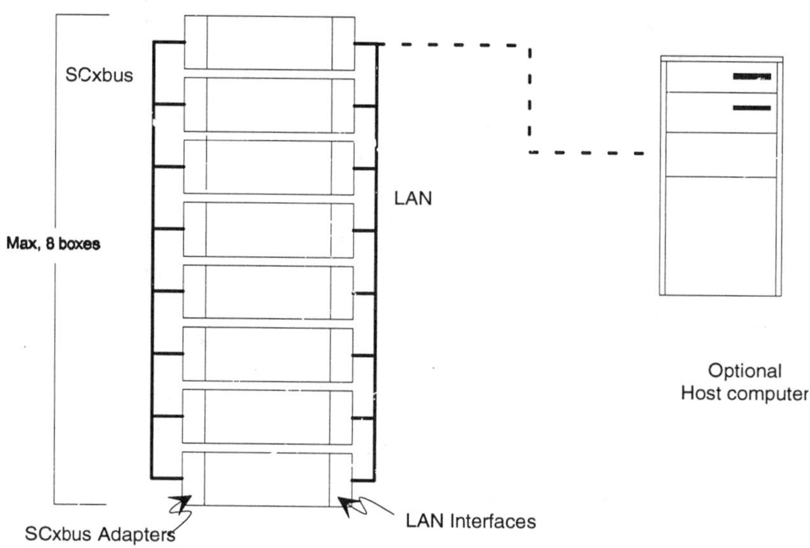

SCxbus Signal Description

Message Bus Interface

SCxSS. SCxbus Supervisory Signal. This line is a serial data line and must be driven with open collector drivers. Data is shifted onto this line with a 2.048 Mbps clock, referred to in this reference as SCxSS clock. SCxSS clock IS NOT available from the SCxbus, and must be generated on every SCxbus adapter.

Data Bus Interface

Symbol	Description	Explanation
SCxCLK	SCxbus system clock	This signal is the SCxbus equivalent of the SCbus signal SCLK, and is driven by the SCxbus clock master. Data transitions occur on the rising edge of SCxCLK. The clock frequency is specified to be 4.096 MHz.
SCxCLKx2*	SCxbus system clock-by-two	This signal is the SCxbus equivalent of the SCbus signal SCLKx2*, and is driven by the SCxbus clock master. It is synchronous to, and twice the frequency of, SCxCLK. Transitions of SCxCLK occur on the falling edge of SCxCLKx2*.
SCxFSYNC *	SCxbus frame sync	This signal is the SCxbus equivalent of the SCbus signal FSYNC*. This active low signal is driven by the SCxbus clock master, and is used to indicate the start of a data frame. SCxFSYNC* goes low for one SCxSCLKx2*-period, one half bit-period prior to the first bit of the first time slot of the frame. The period of SCxFSYNC* is 125ms.
SCxSD [0..15]	Serial Data	These sixteen high lines are driven by individual SCxbus adapters on a time slot by time slot basis. Any one SCxbus adapter may drive any line for any time slot period. The number of time slots in a SCxbus frame is fixed at 64.
SCxCFAIL	SCxbus Clock Fail	This signal is the SCxbus equivalent of the SCbus signal CLKFAIL. This signal is driven by the clock master, and must be driven with open collector drivers. During normal operation of the system clocks SCxCFAIL is held low. If the clock master detects the failure of, or the loss of Synchronization within, the clock source it releases SCxCFAIL, which is then pulled high. SCxCFAIL is used to initiate the reassignment of the role of clock-master from one board to another.

SCxbus Operation

SCxbus Redundancy

The provision of multiple SCxbus adapters in a node may increase the reliability of the system as a whole. When a SCxbus adapter is declared unserviceable all calls on that SCxbus should be terminated. The retention of the call state and a clean transfer to an alternative SCxbus are desirable but not compulsory.

MESSAGE BUS TIMING

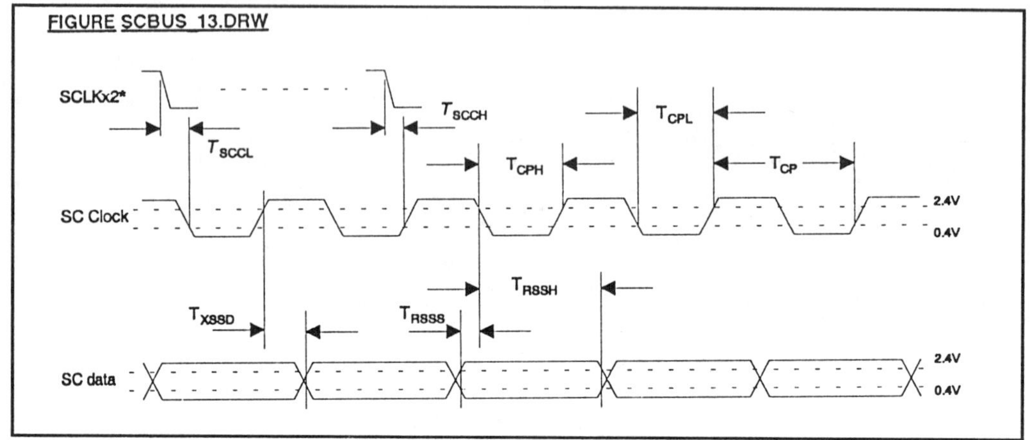

Symbol	Descriptions	Min	Typical	Max	Unit
T sccl	SCLKx2* low to MC clock low delay			25	ns
T scch	SCLKx2* low to MC clock high delay			25	ns
T cph	Clock period high at 2.048 MHz	237	244	252	ns
T cpl	Clock period low at 2.048 MHz	237	244	252	ns
T cp	Clock period at 2.048 MHz	481	488	496	ns
T xmxd	Output data delay - from SCxMC going high to valid output on the bus. Specified for transmitter	25		65	ns
T rmxs	Input data set up time - from valid data to SCxMC going high. Specified for receiver	268			ns
T rmxh	Input data hold time - from SCxMC going high until data change. Specified for receiver	10			ns
T xmxz	Output disable delay - from SCxMC going high to output disabled. Specified for transmitter	15		25	ns

Data Bus

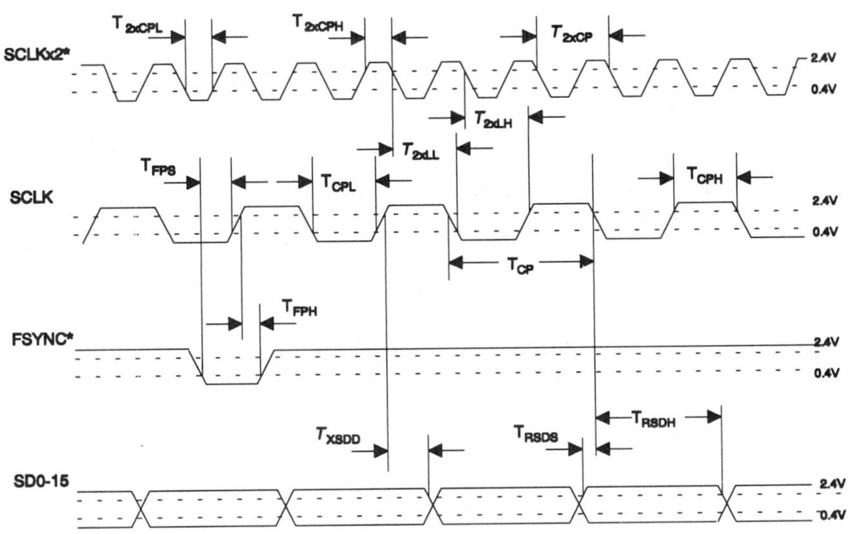

DATA BUS TIMING TABLE SCLK = 2.048 MHz

Symbol	Descriptions	Min	Typical	Max	Unit
T 2xcph	Clock period high at 4.096 MHz	115	122	130	ns
T 2xcpl	Clock period low at 4.096 MHz	115	122	130	ns
T 2xcp	Clock period at 4.096 MHz	237	244	252	ns
T cph	Clock period high at 2.048 MHz	237	244	252	ns
T cpl	Clock period low at 2.048 MHz	237	244	252	ns
T cp	Clock period at 2.048 MHz	481	488	496	ns
T x2ll	4 MHz clock low to 2 MHz clock low delay.			20	ns
T x2lh	4 MHz clock low to 2 MHz clock high delay.			20	ns
T fpd	Frame pulse delay - from SCxCLKx2* going high to valid output on the bus. Specified for transmitter.	10		30	ns
T fps	Frame pulse set-up time - from valid SCxFSYNC* to rising edge of SCxCLK. Specified for receiver.	30			ns
T fph	Frame pulse hold - from rising edge of SCxCLK to valid SCxFSYNC*. Specified for receiver.	10			ns

T xsxd	Output data delay - from SCxCLK going high to valid output on the bus. Specified for transmitter	25		65	ns
T rsxs	Input data set up time - from valid data to SCxCLK going high. Specified for receiver.	268			ns
T rsxh	Input data hold time - valid data after SCxCLK going high. Specified for receiver.	10			ns
T xsxz	Output disable delay - from SCxCLK going high to output disabled. Specified for transmitter	15		25	ns

DATA BUS TIMING TABLE SCLK = 4.096 MHz

Symbol	Descriptions	Min	Typical	Ma x	Unit
T 2xcph	Clock period high at 8.192 MHz	54	61	69	ns
T 2xcpl	Clock period low at 8.192 MHz	54	61	69	ns
T 2xcp	Clock period at 8.192 MHz	115	122	130	ns
T cph	Clock period high at 4.096 MHz	115	122	130	ns
T cpl	Clock period low at 4.096 MHz	115	122	130	ns
T cp	Clock period at 4.096 MHz	237	244	252	ns
T x2ll	8 MHz clock low to 4 MHz clock low delay.			20	ns
T x2lh	8 MHz clock low to 4 MHz clock high delay.			20	ns
T fpd	Frame pulse delay - from SCxCLKx2* going high to valid output on the bus. Specified for transmitter.	10		30	ns
T fps	Frame pulse set-up time - from valid SCxFSYNC* to rising edge of SCxCLK. Specified for receiver.	15			ns
T fph	Frame pulse hold - from rising edge of SCxCLK to valid SCxFSYNC*. Specified for receiver.	10			ns
T xsxd	Output data delay - from SCxCLK going high to valid output on the bus. Specified for transmitter	25		65	ns
T rsxs	Input data set up time - from valid data to SCxCLK going high. Specified for receiver.	24			ns
T rsxh	Input data hold time - valid data after SCxCLK going high. Specified for receiver.	10			ns
T xsxz	Output disable delay - from SCxCLK going high to output disabled. Specified for transmitter	15		25	ns

SCxbus Electrical Specifications

Both SCx data and message bus signals, DATA0 to DATA15, SYNC, CLOCK, and CLOCKx2, use RS485 transceiver with termination resistors at both ends of the cable. The typical RS485 transceivers are TI 75ALS171 or National DS3695/3696 with individual driver disable capability.

SCxbus Mechanical Specifications

SCxbus adapters on the SCxbus are daisy chained together using a common cable or a T-connector. Both ends of the cable are terminated per the RS485 requirement. All signals are common between all devices on the SCxbus. Differential transceivers are used.

Cable and Connector

A 25 signal twisted-pair cable is used. A flat cable can be used in in-cabinet applications, and a shielded twisted round cable can be used in remote applications. The maximum total cable length is 50 feet. A stub length of no more than 3 inches is allowed off the main connector of the SCxbus adapter.

SCxbus Connector Pin out

Pin #	SCxbus Sig.	Pin #	SCxbus Sig.	Description
1	SCxSCLKx2* (A)	26	SCxSCLKx2* (B)	Twice the frequency of SCLK
2	GND	27	GND	Connected to PCB
3	SCxSCLK (A)	28	SCxSCLK (B)	System clock. Driven by clock master.
4	SCxFSYNC*	29	SCxFSYNC*	
5	SCxCFAIL (A)	30	SCxCFAIL (B)	Frame sync pulse. Driven by the clock master.
6	SCxSD0 (A)	31	SCxSD0 (B)	Bidirectional serial data stream # 1.
7	GND	32	GND	Connected to PCB
8	SCxSD1 (A)	33	SCxSD1 (B)	Bidirectional serial data stream # 2.
9	SCxSD2 (A)	34	SCxSD2 (B)	Bidirectional serial data stream # 3.
10	SCxSD3 (A)	35	SCxSD3 (B)	Bidirectional serial data stream # 4.

11	SCxSD4 (A)	36	SCxSD4 (B)	Bidirectional serial data stream # 5.
12	SCxSD5 (A)	37	SCxSD5 (B)	Bi-directional serial data stream # 6.
13	SCxSD6 (A)	38	SCxSD6 (B)	Bi-directional serial data stream # 7.
14	GND	39	GND	Connected to PCB
15	SCxSD7 (A)	40	SCxSD7 (B)	Bi-directional serial data stream # 8.
16	SCxSD8 (A)	41	SCxSD8 (B)	Bidirectional serial data stream # 9.
17	SCxSD9 (A)	42	SCxSD9 (B)	Bidirectional serial data stream # 10.
18	SCxSD10 (A)	43	SCxSD10 (B)	Bidirectional serial data stream # 11.
19	SCxSD11 (A)	44	SCxSD11 (B)	Bidirectional serial data stream # 12.
20	GND	45	GND	Connected to PCB
21	SCxSD12 (A)	46	SCxSD12 (B)	Bidirectional serial data stream # 13.
22	SCxSD13 (A)	47	SCxSD13 (B)	Bi-directional serial data stream # 14.
23	SCxSD14 (A)	48	SCxSD14 (B)	Bi-directional serial data stream # 15.
24	SCxSD15 (A)	49	SCxSD15 (B)	Bi-directional serial data stream # 16.
25	SCxMC (A)	50	SCxMC (B)	Message bus

Device Firmware Services and Messages (SCmessage)

SCmessage Overview

SCmessage provides OSI services, resource management, clocking, and device configuration functions for SCSA devices. It supports interfaces to the SCbus hardware, to application firmware, and to the host interface. Application firmware includes Logical Device Firmware, and algorithm firmware.

FRIMWARE.DRW

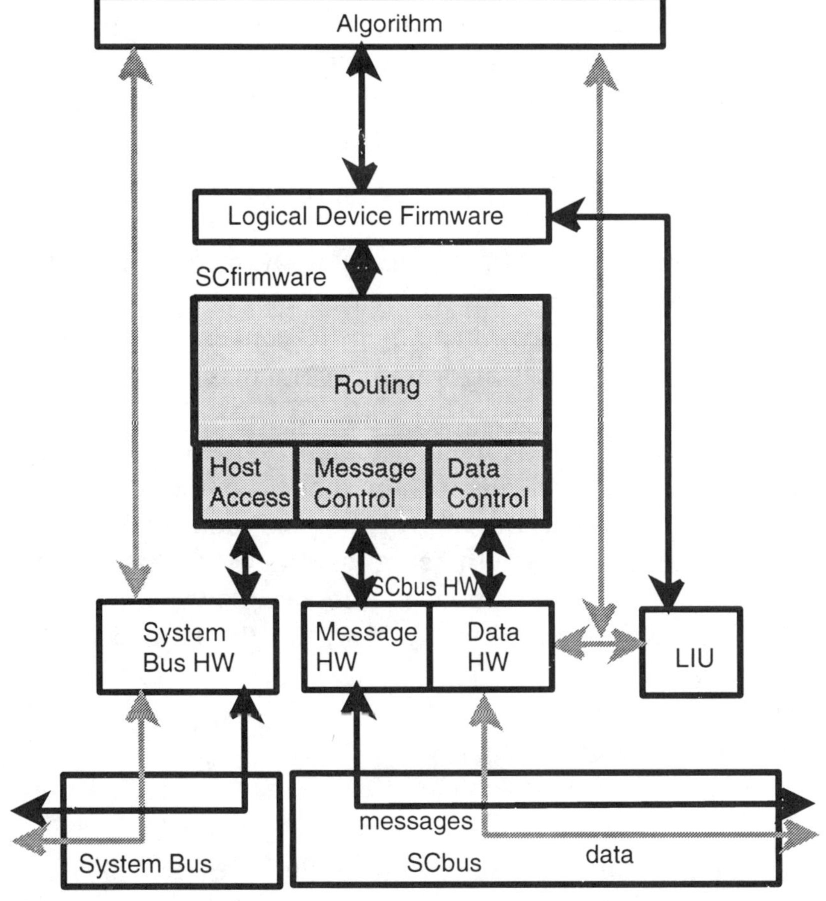

Figure **(FIRMWARE.DRW)** provides a cut-away view of that portion of SCSA which concentrates on firmware at the Call Processing Resource level. This is the layer that sits directly on top of the SCbus and directly below the Host interface and the SCdpi

SCmessage Functions

SCmessage provides OSI Data Link, Network and Transport Layer functions, as well as resource management, clocking, and system configuration support. It supports interfaces to the SCbus hardware, logical device-specific firmware, and the host interface.

It creates a universal port interface to the host computer, providing device location independence and technology integration support. In addition, SCmessage defines a standard logical device interface, and uses a standard communication protocol between processors, boards, and system nodes.

SCmessage Benefits

SCmessage enables the integration of technology-specific boards from different vendors. This is the case, because, the conventions in this specification allow vendors of core technologies to concentrate on their area of expertise, which is in many cases algorithm development.

SCmessage provides SCbus timeslot management and switching control, which helps to assure retention of calls without the loss of clock source. The firmware level is also a point of control for the detection of system configuration changes. The key benefit being that much of what would have been handled by the host application is pressed down into the core technology, thus freeing the application to work on the uniqueness of the user need.

SCmessage Functional Description

SCmessage consists of operating system and hardware dependent control and interface code running on the signal processing device CPU. The SCmessage processes and executes commands generated by the host CPU and the logical device firmware running on the device CPU.

SCmessage consists of the following components.

Component	Description
Routing firmware	The firmware which controls the access to the data and message networks, performs the high level network switching functions, controls access to the host interface, and implements the Logical Device interface
Host access firmware	The firmware which manages the board's host interface, and implements the host interface protocols.
Message bus access firmware	The firmware which manages the message bus interface, and implements the message bus protocols.
Data bus control firmware	The firmware which controls the data bus interface.
Device Services firmware	General configuration management functions and clock fallback support.

Routing Firmware Functional Description

Routing firmware module conforms to OSI Transport and Network Layer standards, allowing for easy substitution, addition or elimination of any Data Link firmware, and providing a standardized interface to Logical Device Firmware.

The routing firmware views all boards in a system (in-box and out-of-box) as well as the Host Computer as part of a single network. It supports the allocation of Logical Devices and the establishment of connections between allocated Logical Devices and the Host Computer. In addition, the firmware routes all messages between Logical Devices and the Host Computer based on a standardized addressing scheme.

The routing module is responsible for maintaining the local configuration data base and initializing the local network configuration. It is also responsible for time slot management, including maintaining a local time slot configuration database, allocating and deallocating time slots, assigning and deassigning time slots, and initializing the local time slot configuration.

Time Slot Model

The SCbus uses nonblocked switching between all time slots. Any time slot on any resource can access any other time slot on the bus. Additionally, one time slot can broadcast to any number of other time slots. All SCbus components have equal time slot switching capability. Figures **(SW_VIEW.DRW) & (TS_VIEW.DRW)** illustrate two instances of the nonblocking switching model.

Host Access Firmware Functional Description

The host access control module is responsible for the operation of the host interface hardware. The host access firmware provides a standard interface between the Routing Firmware and the System Bus (Host Interface) Hardware. It conforms to OSI standards by providing the high-quality Data Link layer necessary for error-free transmission of messages between the Logical Device Firmware, the Host Computer, and the SCbus Hardware.

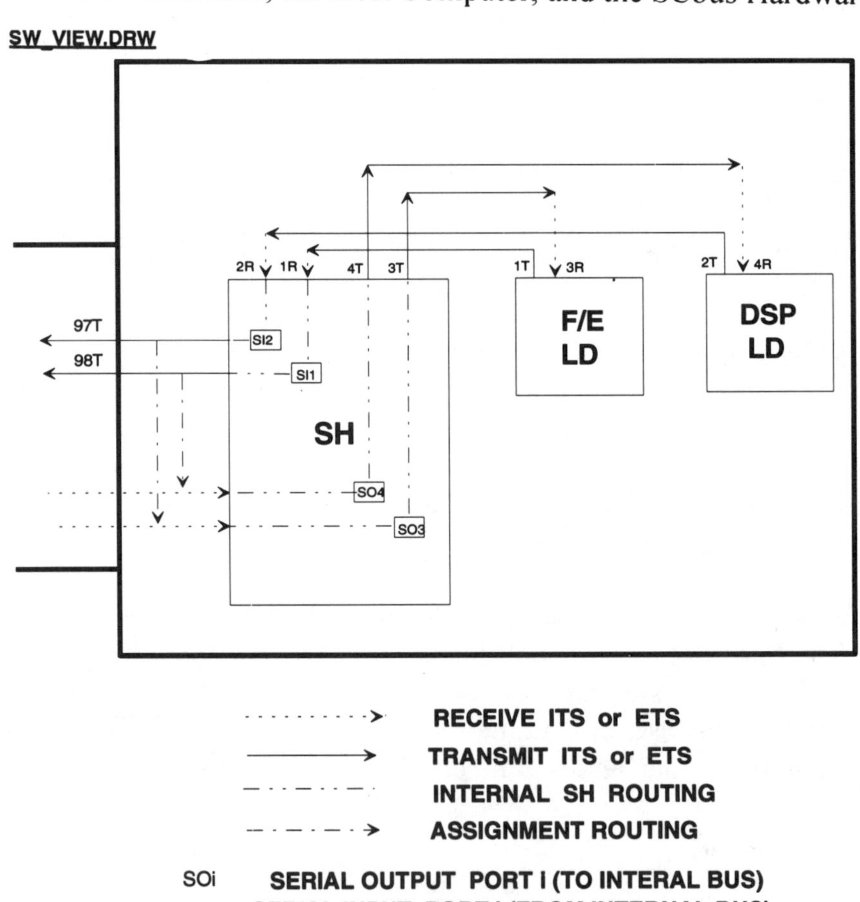

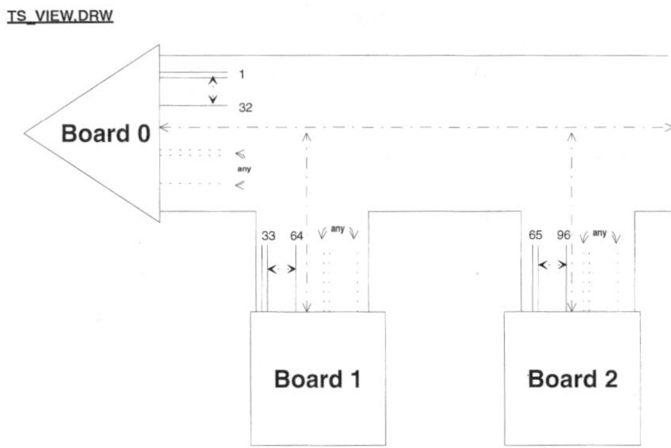

It also performs the protocol conversion required to convert the Host Interface message format to the SCSA standard message format, and enables Technology Developers to easily support SCSA on boards with proprietary Host Interface Hardware by supporting Data Link Firmware created by Technology Developers.

Message Bus Access Firmware Functional Description

The message control module is responsible for implementing the message exchange protocol, including message data framing and error response. In addition, the message control module takes care of the physical transfer of messages to and from the message bus interface hardware. This includes controlling the message bus interface hardware.

It conforms to OSI Data Link Layer standards, and provides a standard interface between the Routing Firmware and the Message Bus Hardware. This interface includes an acknowledgment protocol, flow control, and automatic re-transmission on time-out. It also supports the downloading of run-time firmware, eliminating the need for System Bus (Host Interface) Hardware on boards without Bulk Data requirements.

The message frame format is fully compliant with ISO HDLC UI (Unencumbered Information) frame specifications.

Data Bus Control Firmware Functional Description

The data control module is responsible for implementing the physical transfer of data to and from the data bus interface hardware. It also controls the data bus interface hardware. In addition, the data control module provides a standard interface between the routing firmware and the data bus hardware, and supports all data bus switching functions required to support resource management.

Device Services Firmware Functional Description

The device firmware services module provides general configuration management. It is also responsible for the SCSA clock fall back algorithm. As such, the firmware must handle messages being transmitted or received over the message bus. This requires services to the host computer in order to support seamless recovery procedure, and allows the host to select a fallback clock source prior to clock failure. Upon detection of clock failure, the fallback board assumes the functions of clock master.

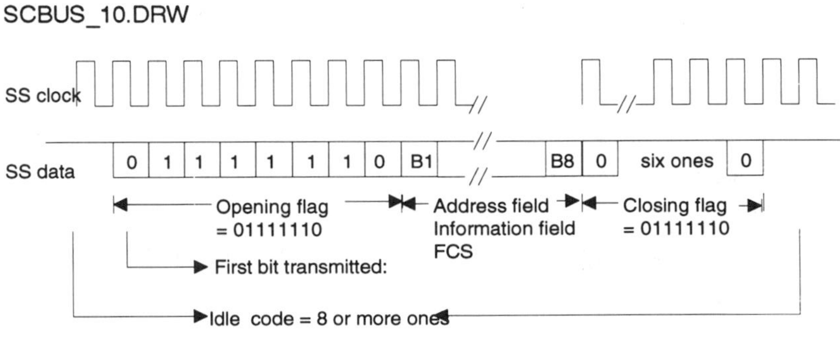

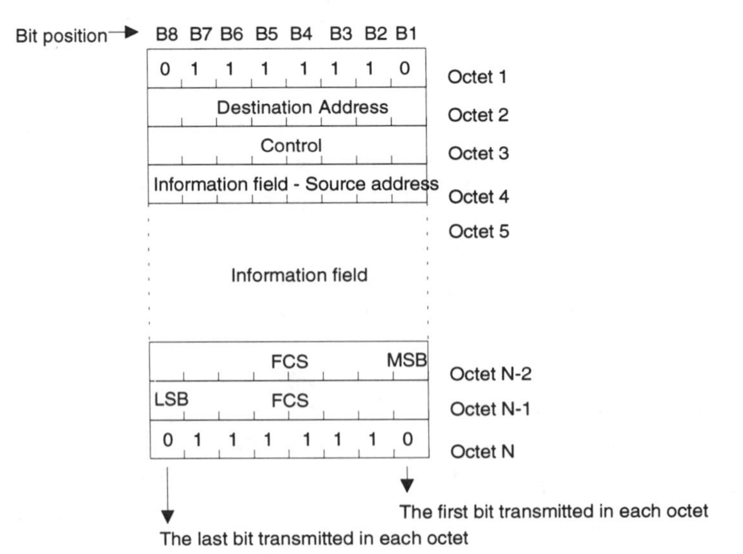

Device Services Firmware handles:

- Configuration Management
- Clock Management

Configuration Management

Configuration Management Description

- Provides services to the Host Computer to support SCSA Configuration Management.
- Supports a Host-originated System Inquiry request to a single board.
- Configuration Management services requests all boards in the system to self-report, collects the responses and replies to the Host.
- Permits the Host Computer to detect the arrival or departure of hot-pluggable nodes with periodic System Inquiries.

Configuration Management Messages

The following configuration management messages are currently proposed.

Host to Configuration Manager

System_Inquiry
Configuration Manager to Configuration Manager:
Board_Inquiry
Board_Information

Clock Management Description

- Provides services to the Host Computer to support a seamless clock recovery procedure.
- Allows Host to select Fallback Clock Source prior to clock failure.
- Upon detection of clock failure, Fallback Board reports failure to Host, becomes clock master, and reports successful mastering to Host.
- Also supports seamless Host-controlled transfer of clock mastering.

Clock Management Messages

The following clock management messages are currently proposed.

Host to Clock Manager

Select_Next_Clock_Master
Deselect_Clock_Master
Clock Manager to Host:
Clock_Master_Failure
Clock_Master_Active

Clock Fallback Procedures

SCbus Procedure for clock fallback

The Host Computer maintains a Clock Fallback List. Each entry of this list contains the Board ID and Clock Source of a possible clock master. The first entry in the list identifies the board to take over as clock master from the boot clock master. The second entry in the list identifies the board and clock source to take over if the first run-time clock master fails.

The last entry in the list identifies the user's last choice for a clock master, and should specify a source that is extremely unlikely to fail (such as an on-board oscillator). If a clock failure occurs when the last entry on list is clock master, the Host Computer will wrap back to the top entry in the list.

The SCmessage supports two Clock Management messages:

Select_Next_Clock_Master
Deselect_Clock_Master.

During the SCSA System Initialization procedure, when it is time to select the first run-time SCbus clock master, the Host Computer sends a Select_Next_Clock_Master message to the Clock Manager firmware identified by the Board ID of first entry in the Clock Fallback List, and specifies the corresponding Clock Source. Subsequently, the Host Computer sends a Deselect_Clock_Master message to the board which is the current boot clock master. This causes the first entry on the Clock Fallback List to become the clock master.

The Host Firmware then selects the second entry in the Clock Fallback List, and sends a Select_Next_Clock_Master message to the Clock Manager firmware identified by the Board ID of that entry, and specifies the corresponding Clock Source. This causes the second entry on the Clock Fallback List to be the "fallback" board, primed to take over as clock master should the current clock master lose its source.

When a clock failure occurs, the "fallback" board reports this occurrence to the Host Computer, which in turn starts a 60ms timer. Should the "fallback" board subsequently report that it successfully took over as clock master before the host's timer expired, the Host Computer cancels its timer.

Should the Host Computer's timer expire before receiving the expected message from the "fallback" board, the host will assume that the "fallback" board's clock source is also bad. In this case, the host selects the next entry in the Clock Fallback List, and sends a Select_Next_Clock_Master message to the designated board, and sends a Deselect_Clock_Master message to the current "fallback" board, in effect telling it to cease its attempt to become clock master. Figure **(SCBUS_31.DRW)** illustrates the logic for the SCSA clock fallback procedure.

SCxbus Procedure for clock fallback

The SCxbus out-of-box procedure relies in part on the SCbus in-box procedure described above. In a multiple node system, there is still only one clock master for the entire system at any one time, except during system initialization. This is true even though there are actually several buses involved: in a two-node system. For example, there is an SCbus in each node and an SCxbus connecting the two nodes by means of a SCxbus board in each node.

Each node has its own local SCbus Clock Fallback List, representing the order of clock mastership for that node. There is also one SCxbus Clock Fallback List, representing the order of clock mastership for the SCxbus. In one node, the SCxbus board (using the SCxbus as its Clock Source) last entry in the local Clock Fallback List. This node is also the first entry in the SCxbus Clock Fallback List (using the SCbus as its Clock Source).

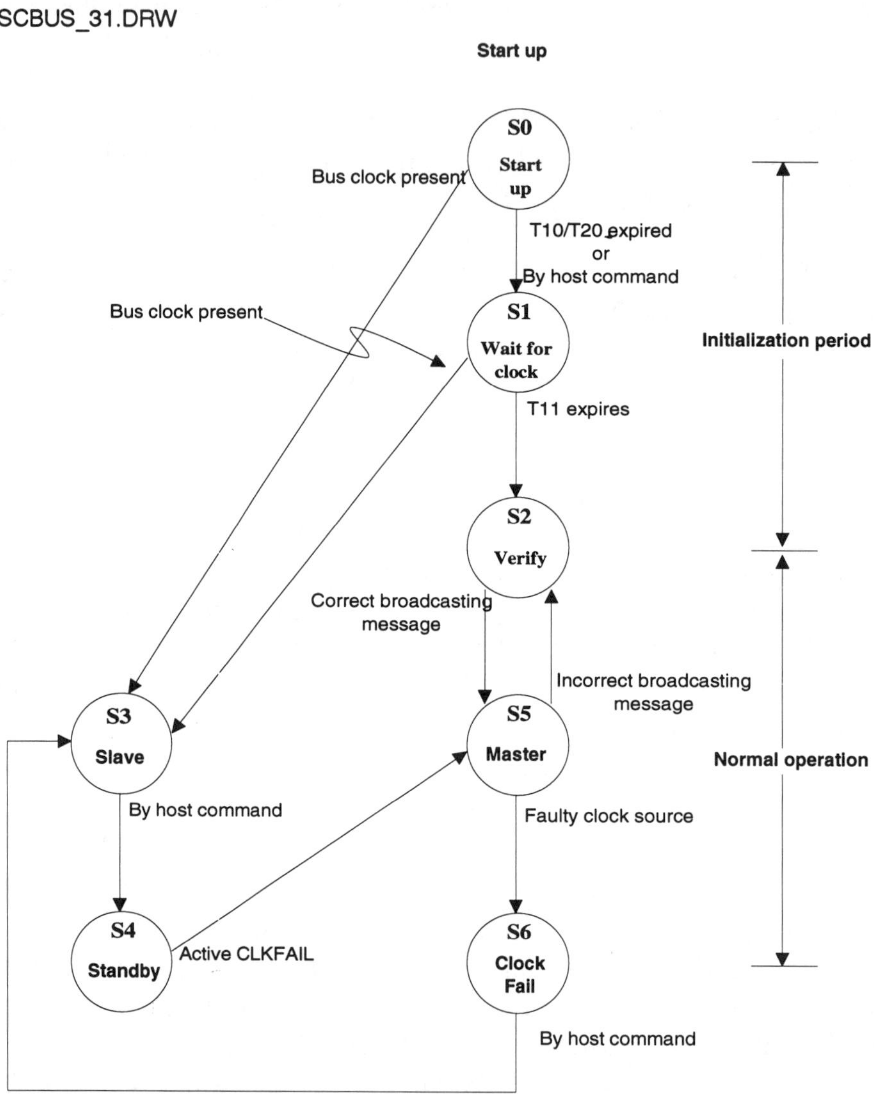

In all other nodes, the SCxbus board is the first entry in each local Clock Fallback List. Each of these other nodes is also represented as additional entries in the SCxbus Clock Fallback List.

When clock failures occur, this scheme permits new boards to be selected as clock master, first within the same node as the current clock master, and then sequentially to the set of boards specified in the Local Fallback List of the node specified as the second entry in the SCxbus Fallback List, and so on.

Device Driver Services and Programming Interface

Device Programming Interface

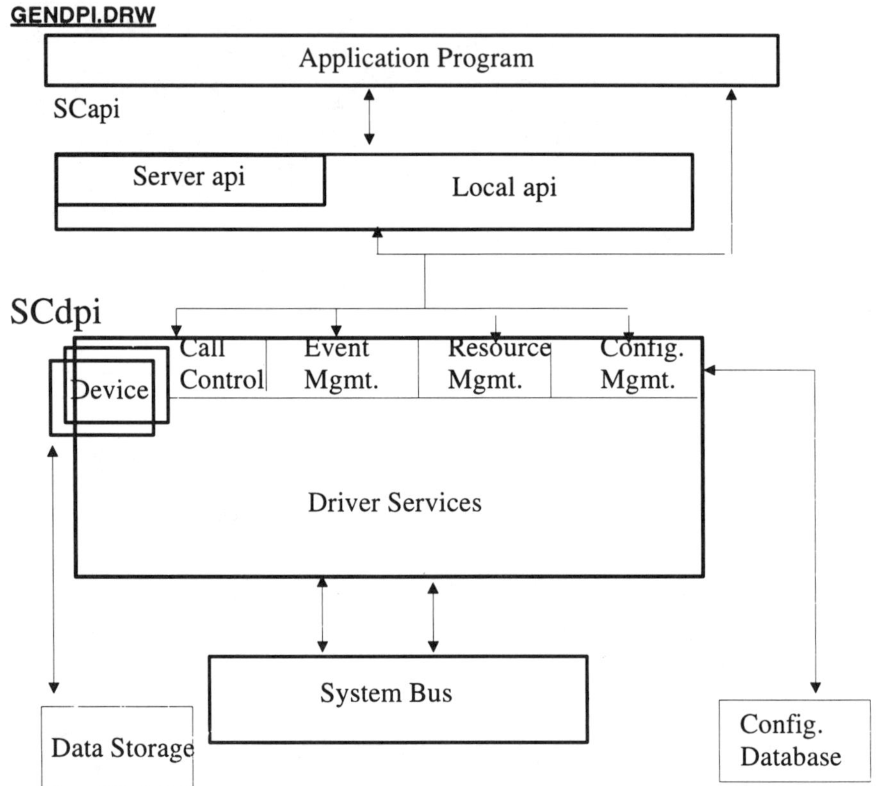

Overview

The SCdpi provides a device-specific interface to each physical device (board, channel, etc.), along with a set of generic services such as configuration management, event management and SCbus management. These services are divided into subparts, and are provided by separate modules within the SCdpi. The interface to each device is provided by a resource module, one per supported device. Facilities exist to add Technology Developer device modules in a specific manner.

The SCdpi is the lowest level in SCSA accessible to an application. All device manipulation must take place through the SCdpi, or through the higher-level Signal Computing API (SCapi). No direct device driver access is provided.

SCSA provides documentation on the access points to a device driver model. The device driver saves developers time in bolting-on their own call processing resources without having to re-do work already done by other manufacturers. A good example would be the Windows operating system. A windows developer who wants to put together a new word processing package, for example, does not have to write 100 printer drivers, because they are already handled by the Windows Shell. The SCdpi works similarly by providing access to common services. These services include:

- Bulk data transfer, buffering and storage management
- Command / reply synchronization
- State maintenance

Module Definitions

The following modules make-up the SCdpi. The tables below provides an explanation for their main functions:

- Signaling
- Resource Allocation
- Configuration Management
- Event Management
- Device Resource Modules
- Configuration Database
- Device Programming Interface

SCdpi Modules

Module Name	Explanation
Signaling	This module provides SCbus signaling functions that are common to all SCSA devices. It performs basic signaling functions, isolating the upper-level software from the physical network.
Resource Allocation	This module provides resource allocation functions that are common to all SCSA devices. It includes mechanisms for assigning resources to ports.
Configuration Management	This module provides common device configuration management functions. It provides functions for querying and updating a system-wide configuration database containing all information about all SCSA devices that are present.
Event Management	This module provides common event handling functions. These are used to enable, disable and retrieve asynchronous event messages from physical SCSA devices.
Device Resource Modules	Each device module provides a device-specific set of functions for manipulating each physical device. While each device is different in the functions it provides, all resource modules share data structures where possible, and conform to a set of standards for their interface specification (function naming, argument layout, etc.).
Configuration Database	A Configuration Database is maintained to support Initialization and configuration of Signal Computing Systems. It contains information such as device names, addresses, status, and resource interface characteristics.
Device Programming Interface	This interface provides device-level programming interfaces and utilities. It includes technology-dependent initialization utilities and downloaders, configuration and event management, and resource allocation services. It also establishes naming and syntax requirements to supports device-specific functions.

SCdpi Definition of Terms

Resource Class, Resource Class Identifier (Class ID)

A *Resource Class* is a particular call processing technology that allows some call processing operation to occur. Examples are voice store and forward, voice recognition, fax send and receive, etc.

A *Resource Class Identifier* (Class ID) is a unique integer identifying a Resource Class. That is, the Class ID identifies some technology type. It is guaranteed that this integer is the same across all SCSA systems, and will always identify the same technology.

Resource Unit (Resource), Unit ID

A *Resource Unit*, or *Resource*, is the smallest addressable and operational part of an instance of a Resource Class. Resource Classes are insatntiated as collections of Resource Units. It is the unit of parallelism for the Resource Class, and the most atomic entity for a resource command.

A *Unit ID* is a unique integer identifying a particular Resource Unit within a Resource Class and Board.

Board, Board ID

A *Board* is an addressable system component that contains Resources. Within a Board, Resources are grouped by Resource Class, and within that Class, are identified by Unit ID. A Board may contain Resources of many different Classes. Thus, within any given Board, the Class ID and Unit ID uniquely identify a Resource Unit. Boards are generally hardware components that are connected together within a Node using the SCbus.

A *Board Identifier* (Board ID) is a unique integer identifying a Board on an SCbus, i.e. within any given Node.

Node, Node Identifier (Node ID)

A *Node* is an addressable system component which contains Boards connected by an SCbus, and is connected to other Nodes via an SCxbus.
A *Node Identifier* (Node ID) is a unique integer identifying each Node in a system.

SCSA System (System)

A *SCSA System*, or *System*, is a collection of Nodes. Each Node is connected to others by an SCxbus in a multi-drop configuration. Multiple parallel SCxbuses may be used for increased bandwidth.

Resource Pool, Pool Identifier

A *Resource Pool*, or *Pool*, is a System-wide, dynamic grouping of particular Resource Units of a Resource Class. Pools are defined at system configuration time. They allow Resources from a Class to be sub-divided into addressable groups, mostly for the purposes of resource allocation.

A *Pool Identifier* is a unique integer identifying a Pool within a System. Resource Address.

A Resource Address is a 5-tuple that identifies a Resource. It is composed of:

- Class ID
- Pool ID
- Board ID
- Node ID
- Unit ID

A Resource Address can identify any Resource within a System. Since Resources may be assigned to several Pools, each Resource may have several Addresses.

A special Pool ID, RA_GLOBAL_POOLID, is always defined, and contains all Resources in a System. This Pool ID is used to ensure that all Resources have at least one Address. Specifying this Pool ID causes the Pool ID to be ignored, reducing the address to a 4-tuple of Class, Node, Board, and Unit ID. This 4-tuple can uniquely identify a single Resource in a System.

In certain situations, a Resource Address may specify a set of resources, rather than just one. This is done by allowing any of the Board, Node and Unit ID fields to contain a wildcard value that matches any possible value in the System. This is primarily used to define a set of Resources from which to do allocation.

In this text, Resource Addresses are written as follows:

C:P:N:B:U

where **C** is the class ID, **P** the Pool ID, **N** the Node ID, **B** the Board ID, and **U** the Unit ID. These may either be symbolic or actual values.

Resource Group, Resource Group Identifier (Group ID)

A *Resource Group*, or *Group*, is a dynamically formed grouping of Resource Units that can be made to work together as a single, universal device. Resource Units grouped in this way may be treated as a single entity; individual resources within the group are addressed using a special short form of address (a Member ID).

A *Group Identifier* (Group ID) is a integer identifying a Group in a system. It is created and assigned when a Group is created. It is guaranteed to be unique to a System, and uniquely identifies a Group within a System.

Group Member (Member), Member Identifier (Member ID)

A Group Member is any Resource Unit that has been attached to a Group. Such a resource is said to be a member of the Group to which it is attached.

A *Member Identifier* (Member ID) is an integer that uniquely identifies any attached resource in a system. It is created and assigned when a Resource is attached to a Group. It is possible to derive a Group ID from a Member ID.

Device, Device ID

A Device is either a Group or a Member.

A Device ID identifies a Device; that is, it is either a Group ID or a Member ID.

Device Handle

A *Device Handle* is used by the SCdpi to identify a Device to operate on. It will therefore contain a Device ID, as well as other, implementation-specific information about a Device.

Programming Model Definition

The job of the SCdpi is to manage resource allocation, allow Resources to be grouped together and treated as a single device, and to manage data connections between these Resources.

Within an application using SCdpi, call processing is done by creating Groups of Resources, specifying and manipulating data connections between Members and Groups, and invoking operations on Groups, which form virtual universal device devices.

Groups are the central focus for the SCdpi. They are formed explicitly using SCdpi functions, are dynamic in nature, and may be formed, destroyed and altered at will. The cost of these operations is very low. Groups are almost always associated with a call, and represent the group of Resource Units needed at any particular moment to achieve some call processing aim.

For instance, an application may want to switch between using voice output, text to speech, and fax output within a single call. This can easily and swiftly be done by creating a Group, attaching the relevant Resource Units, and switching between them as needed.

Groups may have any Resource desired (that is available), and may be thought of as creating virtual universal devices. Note also that this virtual device can be dynamically configured, and may have its technology properties changed at any time, by attaching and detaching Resources.

Forming Resource Groups

Groups are formed by *claiming* a Resource Unit as the initial member. This will allocate the Group ID, and also the first Member ID. Both are returned as Device Handles.

Almost always, this first member will be a Unit representing a network interface of some kind, although for specialized circumstances, this does not have to be the case. However, since call processing involves working with a call and caller, a network interface is usually needed throughout the transaction, and once created, would not be removed from the Group. There are however, some notable exceptions to this, such as when sharing a resource between several groups. In this case, the shared resource would be placed in a separate group of its own, and manually connected to members in other groups.

Once a Group has been formed, other Resource Units may be added (using the *attach* operation) to perform call processing operations. For instance, a voice Resource may be added to do voice store and forward, a text to speech resource may be added to read some text, etc. The Resource Unit to be attached is specified using a Resource Address.

Groups may not contain more than one member of any given resource type (Class ID). Claiming or attaching a Resource Unit marks it as allocated, and makes it unavailable for other claims and attaches for as long as it remains allocated to the group.

Operating on Resource Groups

At the SCdpi level, all operations are invoked by function calls on a Device Handle. Since this can refer to either a Group or a Member, all SCdpi functions can operate on either a Group or a Member. There are defined general rules for how SCdpi functions differentiate these two types of Devices. In most cases, a Group will be used, and the operation functions are able to decide which Group Member to apply an operation to. It is this property that allows Groups to be treated as a single entity, and operations applied to them as a whole, without having to specify which Member to operate on.

When a function is invoked on a Group, the function tries to determine which Member of the Group has the correct Resource ID implied by the operation. If this member exists, the operation applied to it. If not, an error is returned.

Conversely, when a function is invoked on a Member, the operation defined by the function is invoked on that Member. If the operation is inappropriate for the Resource (i.e. it is the wrong type), an error is returned.

In some cases, an application may require that an operation be directed to a particular Member. In this case, a Device Handle that refers to the Member (rather than the Group) should be used. It is the programmer's responsibility to provide this handle to the function when a specific device is required. Functions which are special cases in this respect are noted in the function descriptions below.

Switching Model

The SCSA switching model is a set of rules and functions that allow simulated many-to-one data connections between resources. These connections are primarily intended between members of the same group, but inter-group connections are also supported to a limited extent.

Because of the underlying switching matrix, SCbus resources can only accept input from one source at a time; using several resources within a group therefore requires constant switching between them as they are activated. This switching management is provided by the SCSA switching model. The model defines the ability to make virtual many-to-one connections between members, and ensures that these connections are switched appropriately as the connected resources are used. This happens transparently to the user of the SCdpi, and provides the all of the functionality of a set of simultaneous many-to-one connections.

Since many of these connections are fixed, in the sense that certain resources will always connect to certain others, the model also embodies the idea of being able to set up a fixed set of data relationships, which can be selectively overridden as required. This allows fixed patterns of connections to be set up once (and maintained by the SCdpi), while allowing these fixed patterns to be modified selectively as a situation demands.

Groups are the main focus for this switching model, since resources that need to be connected will almost always be part of a group solving a common problem. At times, it is required to connect between members of different groups, allowing a form of group connection. This is also supported by the model.

Permanent Connections

Permanent connections are data paths maintained and managed by the SCdpi, such that they will always be in effect when the related resources are active. They may only be set up when a resource is attached to a group.

As part of the attach operation, a connection target for the member may be specified. If this is done, the SCdpi will ensure that the target and source members are connected whenever the source is active. Thus, for situations requiring a fixed set of relationships, the attaching commands may specify this requirement, and the SCdpi will ensure that these relationships are maintained without further work from the application. Attach relationships must be many to one, and so can only form a tree structure.

Temporary Connections

The above scheme, while being excellent for managing permanent, fixed connections within groups, can be too restrictive for some applications, and does not allow for inter-group connections. To solve this problem, the SCdpi allows these fixed relationships to be temporarily and explicitly overridden by the application.

This is done with the *connect* operation. This connects a source to a target, overriding any permanent connection, until overridden in turn by another connect, or removed by an unconnect operation. At this point, any previous attach connection (permanent connection) is restored.

Successive explicit connections override each other, and do not nest. That is, an explicit connect will override a previous one, and will cause the first one to be lost. Only connections specified at attach time are restored when an unconnect operation is performed.

Event Handling

Groups also form the prime mechanism for handling events. Events from Members of a Group are sent to the Group, and may be retrieved by using functions from the Event Handling section (see below) on the Group Device. These events contain an indication of the Device Handle of the Member from which they originated, an event type, and variable length event data. These may all be retrieved using Event Handling functions.

Event handling may be either synchronous or asynchronous. In synchronous mode, the application waits explicitly for one event, and finishes processing that event before performing other tasks. This is sufficient for the simplest applications. In asynchronous mode, the application processes events as they occur, without knowing when or in what order they will occur. Many SCdpi functions take a *mode* argument specifying whether the function is to run in synchronous or asynchronous mode.

The SCdpi model supports synchronous event handling and three types of asynchronous event handling, as illustrated in the following code fragments.

Synchronous Mode

The application chooses a specific event to wait for, and processes that event in its main application code when it occurs.

```
Main() {

    waitcall()
    play()
    record()

    }
```

Asynchronous Polled Mode

The main application code waits for any event to occur, and processes that event according to its type, what device it occurred on, and the state of the device at the time. This model is convenient for applications in which not many kinds of events will occur.

```
process_event() {

    switch ( Next_Action( current_state, event ) ) {
    case PLAY:
        async_play()
        return
    case RECORD:
        async_record()
        return
        .
        .
        .
}

Main() {

    while(Not_Aborted) {
        waitevent()
        process_event()
    }
}
```

Asynchronous Callback Mode

The application specifies an event handler for each kind of event that is expected to occur. This model is more structured than the polled model, and is suitable for more complex applications in which the polled model might be expected to present code maintenance difficulties.

```
void event_handler_1() {
    /* Process my events */
        .
        .
        .
}
    .
    .
```

```
.
event_handler_N() {
    /* Process my events */
        .
        .
        .
}

Main() {

    enable_event_handler(event_handler1);
        .
        .
        .
    enable_event_handler(event_handlerN);
    while(Not_Aborted) {
        waitevent()
    }
}
```

Asynchronous Interrupt Mode

The application specifies event handlers as for the callback mode, but performs other tasks in the foreground; these foreground tasks are interrupted randomly whenever an event occurs.

```
void event_handler_1() {
    /* Process my events */
        .
        .
        .
}
    .
    .
    .
event_handlerN() {
    /* Process my events */
        .
        .
```

```
}

Main() {

     enable_event_handler(event_handler1);
          .
          .
          .
     enable_event_handler(event_handlerN);
     /* Execute main code */
          .
          .
          .
}
```

Generic Driver Services

All bulk data handling between the SCdpi functions and the SCSA hardware is performed by a single device driver services module. This module maintains configuration information for all physical devices and executes all synchronization and routing operations. Application programs do not issue commands directly to the driver services module; these are all encapsulated in SCdpi functions. The extensibility and the technology-independent nature of the SCdpi (including the driver services module) have been designed to ensure support for future communications technologies.

SCdpi Components

The SCdpi is divided into a section of attribute functions, four sections of common functions, and several resource-specific sections. The four common sections are Call Control, Event Management, Resource Management and Configuration Management.

Attribute Functions

Attribute functions are functions that return information about Devices or the state of the system. They have a specific format and argument syntax, and are designed to provide an extensible mechanism for returning state and error information. They always take a Device Handle as an argument, and return a single value, which may be a structure. They are named in all upper-case letters, with the prefix **AT_**.

Resource specific modules may also define attributes. These are named with the same **AT_** prefix, but with the addition of a three-letter resource identifier; for example, **ATVOX_** is the attribute prefix for the Voice Store and Forward module.

Call Control

Functions in the Call Control section are concerned with setting up, controlling and tearing down calls. This section provides functions to perform these operations in a network and protocol independent manner, including the collection of any caller and calling ID digits, the generation of special tones and timing, and other network control functions.

Resource Management

Functions in the Resource Management section are concerned with allocating Resource Units and Groups, attaching and detaching Units to and from Groups, and connecting and switching Units.

Event Management

Functions in the Event Management section manage and collect events from Resource Units. These functions may be used to poll for events, set event handling functions to be called when an event occurs, and enable and disable event generation.

Configuration Management

Functions in the Configuration Management section are concerned with querying and updating the Configuration Database. This includes the translation of symbolic Resource Unit names to Resource Addresses that can be used by other SCdpi functions.

Common Device Support

These functions provide ancillary support functions, such as maintaining terminating condition tables, and canceling commands in progress, which are common to all devices.

SCdpi Function Summary

Attribute Functions

AT_CALLRESULT( )	get outgoing call result
AT_CALLER_ADDR( )	get caller identification
AT_CALLED_ADDR( )	get called identification
AT_TERMMASK( )	get termination bit mask
AT_TRCOUNT( )	get transfer count for last operation
AT_ERRMSGP( )	print string for last error that occurred
AT_RESOURCES( )	return resources attached to a device
AT_LASTERR( )	return last error that occurred
AT_CALLSTATE( )	return the current call state

Call Control Functions

SCcon_makecall( )	make a call
SCcon_ackcall( )	acknowledge/reject a call
SCcon_setlinestate( )	set line service statc
SCcon_getlinestate( )	get line service state
SCcon_teardowncall( )	tear down a call

Resource Management Functions

SCrsc_claim( )	claim a Resource Unit and create a Group
SCrsc_release( )	release a Group and its attached resources
SCrsc_attach( )	attach a Resource Unit to a Group
SCrsc_detach( )	detach a Resource Unit from a Group
SCrsc_connect( )	Temporarily connect members
SCrsc_unconnect( )	Undo a connect operation
SCrsc_getparm( )	return a Resource Unit parameter value
SCrsc_setparm( )	set a Resource Unit parameter value

Event Management Functions

SCevt_enbhdlr( )	enable an event handler
SCevt_dishdlr( )	disable a handler
SCevt_waitevt( )	get next event
SCevt_getevtdev( )	get device handle of current event. source
SCevt_getevttype( )	get event type for current event
SCevt_getevtlen( )	get length of data associated with the current event
SCevt_getevtdatap( )	get pointer to data associated with the current event.

Configuration Management Functions

TBD

Common Device Support Functions

SCdev_clrtpt( )	clear a Termination Parameter Table entry
SCdev_stop( )	stop a command

Voice Store and Forward Functions

ATVOX_BUFDIG( )	get number of digits in buffer
SCvox_clrdigbuf( )	clear the firmware digit buffer
SCvox_getdig( )	get a digit string
SCvox_play( )	play voice data
SCvox_rec( )	record voice data
SCvox_playf( )	play a voice file
SCvox_recf( )	record a voice file

SCdpi Type Definitions

Basic Types

Synopsis

```
typedef CHAR        /* 8-bit signed */
typedef SHORT       /* 16-bit signed */
typedef LONG        /* 32-bit signed */
typedef UCHAR       /* 8-bit unsigned */
typedef USHORT      /* 16-bit unsigned */
typedef ULONG       /* 32-bit unsigned */

typedef UnitID      UCHAR;
typedef BoardID     UCHAR;
typedef NodeID      UCHAR;
typedef PoolID      UCHAR;
typedef ResourceID  UCHAR;
```

Description

In the following sections, these basic types are assumed to be defined as shown above.

Type SC_DEVHNDL

Synopsis

```
typedef sc_dev { ... } SC_DEV;

typedef SC_DEV *SC_DEVHNDL;
```

Description

An SC_DEVHNDL object identifies either a Resource Group or a Resource Group Member. It is created when a Group is formed by SCrsc_claim( ), or when a Resource Unit is added to a group by SCrsc_attach( ). It is passed as the first parameter of all other operation functions to identify the target of the operation.

Type SC_RESADDR

Synopsis

```
typedef struct sc_resaddr SC_RESADDR {
  ResourceID  res;
  PoolID      pool;
  NodeID      node;
  BoardID     board;
  UnitID      unit;
};

#define RA_GLOBAL_POOLID      0      /* Global pool ID */
#define RA_ANYVALUE                  /* Wildcard value */
```

Description

An SC_RESADDR is used to identify a single Resource unit, or a set of Resource Units.

Type SC_TPT

Synopsis

```
typedef struct sc_tpt {
  USHORT  tp_type;     /* Flags describing this entry  */
  USHORT  tp_termno;   /* Termination Parameter number */
  USHORT  tp_length;   /* Length of terminator         */
  USHORT  tp_flags;    /* Parameter attribute flag     */
  USHORT  tp_data;     /* Optional additional data     */
  USHORT  rfu;         /* Reserved                     */
  SC_TPT  *tp_nextp;   /* Pointer to next termination  */
                       /* parameter if IO_LINK set     */
} SC_TPT;
```

Description

An SC_TPT is an entry is a Termination Parameter Table that defines one or more terminating conditions for a command.

Type SC_IOTT

Synopsis

```
typedef struct sc_iott {
  USHORT   io_type;          /* Transfer type */
  USHORT   rfu;              /* Reserved     */
  LONG     io_fhandle;       /* File descriptor */
  char     *io_bufp;         /* Pointer to base memory */
  ULONG    io_offset;        /* File/Buffer offset */
  LONG     io_length;        /* Length of data */
  SC_IOTT *io_nextp;         /* Ptr. to next entry if */
                    /* IO_LINK set */
  SC_IOTT *io_prevp;         /* Ptr. to previous entry */
                    /* (Optional) */
} SC_IOTT;

/* Flag values */

#define IO_CONT  0x01     /* Next entry is contiguous */
                /* (DEFAULT) */
#define IO_LINK  0x02     /* Next entry is linked */
#define IO_EOT   0x04     /* This is the last entry
#define IO_DEV   0x00     /* play from a file */
#define IO_MEM   0x08     /* play from memory */
```

Description

An SC_IOTT is an entry in a I/O Transfer Table that defines the data sources or sinks for a command that transfers bulk data.

Type EVT_HANDLER

Synopsis

```
typedef LONG (*EVT_HANDLER)(...);
```

Description

An EVT_HANDLER is a pointer to an event handling function.

SCdpi Function Definitions

Attribute Function Definitions

AT_CALLRESULT( ) - *get outgoing call result*

Synopsis

USHORT AT_CALLRESULT (SC_DEVHNDL dev);

Description

The attribute function AT_CALLRESULT returns the result of the last outgoing call placed on the device *dev*. If this is a Group, it must have a network interface resource attached. If it is a Member, it must be a network Resource Unit.

AT_CALLER_ADDR( ) - *get caller identification*

Synopsis

const char *AT_CALLER_ADDR (SC_DEVHNDL dev);

Description

The attribute function AT_CALLER_ADDR returns the identity of the caller for the current call on device *dev*. If this is a Group, it must have a network interface resource attached. If it is a Member, it must be a network Resource Unit.

AT_CALLED_ADDR( ) - *get called identification*

Synopsis

const char *AT_CALLED_ADDR (SC_DEVHNDL dev);

Description

The attribute function AT_CALLED_ADDR returns the identity of the address that was called for the current call on device *dev*. If this is a Group, it must have a network interface resource attached. If it is a Member, it must be a network Resource Unit.

AT_TERMMASK( ) - *return termination bit mask*

Synopsis

ULONG AT_TERMMASK (SC_DEVHNDL dev);

Description

This function returns the termination mask for the last operation on the Device *dev*.

AT_TRCOUNT( ) - *get transfer count for last operation*

Synopsis

ULONG AT_TRCOUNT (SC_DEVHNDL dev);

Description

This function returns the number of bytes transferred by the last operation on the Device *dev*.

AT_ERRMSGP( ) - *print string for last error that occurred*

Synopsis

const char *AT_ERRMSGP (SC_DEVHNDL dev);

Description

The attribute function AT_ERRMSGP( ) returns the print string for the last error that occurred on the device *dev*.

AT_LASTERR( ) - *return last error that occurred*

Synopsis

USHORT AT_LASTERR (SC_DEVHNDL dev);

Description

The attribute function AT_LASTERR( ) returns the error number of the last error that occurred on the device *dev*.

AT_CALLSTATE( ) - *return the current call state*

Synopsis

USHORT AT_CALLSTATE (SC_DEVHNDL dev);

Description

The attribute function AT_CALLSTATE( ) returns the current state of the call on the device *dev*.

Call Control Function Definitions

SCcon_makecall( ) - *make a call*

Synopsis

```
USHORT SCcon_makecall (SC_DEVHNDL dev, UCHAR mode,
                char *addr, void *udata,
                USHORT ulen);
```

Description

This function places a call on the Device *dev*. It must be a Group, and have a network Resource Unit as a member. The string *addr* specifies the address to call, and the data block *udata* contains any user-to-user data. This block has length *ulen*. *mode* specifies if the function should be executed synchronously or asynchronously.

SCcon_ackcall( ) - *acknowledge/reject a call*

Synopsis

```
USHORT SCcon_ackcall (SC_DEVHNDL dev, UCHAR mode,
                UCHAR response, void *udata,
                USHORT ulen);
```

Description

This function acknowledges an incoming call on the Device *dev*. The call can either be accepted or rejected; *response* specifies the action required. The block of data at *udata*, of length *ulen*, will be filled with any user-to-user data. *mode* specifies if the function should be executed synchronously or asynchronously.

SCcon_setlinestate( ) - *set line service state*

Synopsis

```
USHORT SCcon_setlinestate (SC_DEVHNDL dev, UCHAR state);
```

Description

This function sets the service state of the telephone line associated with the Group *dev*. This group must contain a network Resource Unit that identifies the line to set. *state* specifies the state to set.

SCcon_getlinestate( ) - *get line service state*

Synopsis

USHORT SCcon_getlinestate (SC_DEVHNDL dev, UCHAR *statep);

Description

This function gets the service state of the telephone line associated with the Group *dev*. This group must contain a network Resource Unit that identifies the line to report on. The state of the line is returned in the variable pointed to by *statep*.

SCcon_teardowncall( ) - *tear down a call*

Synopsis

USHORT SCcon_teardowncall (SC_DEVHNDL dev, UCHAR mode);

Description

This function tears down a call in progress on the Device *dev*. *mode* specifies if the function should be executed synchronously or asynchronously.

Resource Management Function Definitions

SCrsc_claim( ) - *claim a Resource Unit and create a Group*

Synopsis

```
SC_DEVHNDL SCrsc_claim (SC_RESADDR res,
                SC_DEVHNDL *membhandle);
```

Description

This function claims a Resource Unit and creates a Group with that Resource Unit as its first member. *res* is a Resource Address and specifies the set of Resource Units from which to allocate. The function returns the handle of the new Group, and returns the handle of the first member of the group in the variable pointed to by *membhandle*.

The function will fail and return SC_NODEV if the Resource Unit has already been claimed.

SCrsc_release( ) - *release a Group and its attached resources*

Synopsis

```
USHORT SCrsc_release (SC_DEVHNDL dev);
```

Description

This function releases a claimed Resource Unit, and detaches all other attached Resource Units.

SCrsc_attach( ) - *attach a Resource Unit to a Group*

Synopsis

```
SC_DEVHNDL SCrsc_attach (SC_DEVHNDL group, SC_RESADDR res,
                SC_DEVHNDL target);
```

Description

This function allocates and attaches a Resource Unit to a Group. The Unit to be allocated is specified by *res*; it will be attached to the Group specified by *group*. No other resources of the same Class ID as *res* may be attached to *group*.

If the member *target* is supplied, a permanent connection between the added member and *target* is set up. This may be SC_NODEV to inhibit this connection.

SCrsc_detach( ) - *detach a Resource Unit from a Group*

Synopsis

USHORT SCrsc_detach (SC_DEVHNDL memb);

Description

This function detaches a previously attached Unit. *memb* is the handle of the attached Unit. It must refer to a single member, not a Group.

SCrsc_connect( ) - *Temporarily connect members*

Synopsis

USHORT SCrsc_connect (SC_DEVHNDL src, SC_DEVHNDL target);

Description

This function sets up a temporary connection between a source and a listener. *src* specifies the source of the data, and *target* the listener. This connection remains in place, overriding any attach connection, until it is either replaced by another connection involving *src* or *target*, or is removed by a call to **SCrsc_unconnect( )**. The temporary connection is bidirectional if both the source and target are bidirectional Resources; otherwise the connection is unidirectional in the appropriate direction.

SCrsc_unconnect( ) - *Undo a temporary connection*

Synopsis

USHORT SCrsc_unconnect (SC_DEVHNDL src, SC_DEVHNDL target);

Description

This function removes a temporary connection made by **SCrsc_connect( )**, restoring the effect of any attach connection involving *src* and *target.*

SCrsc_getparm( ) - *return a Resource Unit parameter value*

Synopsis

```
USHORT SCrsc_getparm (SC_DEVHNDL dev, ULONG parmID,
            void *valp);
```

Description

This function returns the value of a parameter for the Resource Unit identified by *dev*. This must be a Member ID. The parameter to get is defined by *parmID.*

SCrsc_setparm( ) - *set a Resource Unit parameter value*

Synopsis

```
USHORT SCrsc_setparm (SC_DEVHNDL dev, ULONG parmID,
            void *valp);
```

Description

This function sets the value of a parameter for the Resource Unit identified by *dev*. This must be a Member ID. The parameter to set is defined by *parmID*.

Event Management Function Definitions

SCevt_enblhandler( ) - *enable an event handler*

Synopsis

```
USHORT SCevt_enblhandler (SC_DEVHNDL dev, EventID evt,
                    EVT_HANDLER *handler);
```

Description

This function enables the event handler *handler* for the event type *evt* on the Device *dev*. This device must be a Group.

SCevt_dishandler( ) - *disable a handler*

Synopsis

```
USHORT SCevt_dishandler (SC_DEVHNDL dev, EventID evt,
                    EVT_HANDLER *handler);
```

Description

This function disables the event handler *handler* for the event type *evt* on the Device *dev*. This device must be a Group.

SCevt_waitevt( ) - *get next event*

Synopsis

USHORT SCevt_waitevt (SC_DEVHNDL dev, EventID evt,
    LONG timeout);

Description

This function waits for an event of type *evt* on the Device *dev*. This device must be a Group. All enabled event handles for this event have been called and have returned when this function returns.

SCevt_getevtdev( ) - *get device handle for current event*

Synopsis

SC_DEVHNDL SCevt_getevtdev();

Description

This function returns the Group handle for the current event. It returns SC_NODEV if there is no current event.

SCevt_getevttype( ) - *get event type current event*

Synopsis

EventID SCevt_getevttype();

Description

This function returns the type of the current event. If there is no current event, it returns SC_NOEVENT.

SCevt_getevtlen( ) - ***get length of data associated with the current event***

Synopsis

USHORT SCevt_getevtlen();

Description

This function returns the length of the variable data associated with the current event. If there is no current event, or there was an error retrieving the last event (including a timeout), it returns -1.

SCevt_getevtdatap( ) - ***get pointer to data associated with the current event***

Synopsis

void *SCevt_getevtdatap();

Description

This function returns a pointer to the variable data block associated with the current event. If there is no current event, or there was an error retrieving the last event (including a timeout), it returns a NULL pointer.

Common Device Support Function Definitions

SCdev_clrtpt( ) - ***clear the Termination Parameter Table entry***

Synopsis

USHORT SCdev_clrtpt (SC_TPT *tptp, USHORT size);

Description

This function clears a number of individual SC_TPT structures according to the parameter *size*. Prior to calling this function, the user must set the *tp_type* fields to IO_CONT if the next SC_TPT is contiguous or to IO_EOT for the last SC_TPT. If the *tp_type* field is set to IO_LINK, the user must set *tp_nextp* to point to the next SC_TPT in the chain.

The function clears all fields except the *tp_type*, *tp_prevp* and *tp_nextp* fields of the number of SC_TPT structures specified by *size*. *tptp* is a pointer to the first SC_TPT to be cleared. The function examines the *tp_type* field of the SC_TPT (and the *tp_nextp* field if IO_LINK is set) to access the next SC_TPT. Thus, a combination of contiguous and linked SC_TPT structures may also be used. If *size* is set to zero, the function simply returns. If IO_EOT is encountered in the *tp_type* field before *size* number of SC_TPTs are cleared, an error will be returned.

SCdev_stop( ) - *stop a command*

Synopsis

USHORT SCdev_stop (SC_DEVHNDL dev);

Description

This function stops an operation currently in progress on Device *dev*.

Voice Store and Forward Function Definitions

ATVOX_BUFDIGITS( ) - *return number of digits in buffer*

Synopsis

USHORT ATVOX_BUFDIGITS (SC_DEVHNDL dev);

Description

This function returns the number of digits buffered by the firmware on the Device *dev*.

SCvox_clrdigbuf( ) - ***clear the firmware digit buffer***

Synopsis

```
USHORT SCvox_clrdigbuf (SC_DEVHNDL dev);
```

Description

This function clears the firmware digit buffer for the Device *dev*.

SCvox_getdig( ) - ***get a digit string***

Synopsis

```
USHORT SCvox_getdig (SC_DEVHNDL dev, SC_TPT *tptp,
                VOX_DIGIT *digbufp, UCHAR mode);
```

Description

This function initiates the collection of digits from the firmware. The termination conditions are specified in the SC_TPT structure(s) pointed to by *tptp*. The parameter *digbufp* is a pointer to a digit buffer which is implemented as a VOX_DIGIT structure. The digits collected from firmware (both the digit ASCII value and the digit type) will be transferred to this VOX_DIGIT structure in the user space.

mode defines whether the function execution is synchronous or asynchronous. When run asynchronously, an EV_GETDIG event will be posted when the operation is complete. The application may use the AT_TERMMASK( ) attribute to determine the exact reason for termination.

SCvox_play( ) - ***play voice data***

Synopsis

```
USHORT SCvox_play (SC_DEVHNDL dev, SC_IOTT *iottp,
               SC_TPT *tptp, UCHAR mode);
```

Description

This function plays a sequence of voice data specified by an array or list of SC_IOTT structures pointed to by *iottp*. This data is played until the end of the data is reached, or one of the terminating conditions specified in the array or list of terminating conditions pointed to by *tptp* is satisfied. If *tptp* is NULL then previously set terminating conditions will take effect.

mode defines whether the function execution is synchronous or asynchronous. When run asynchronously, an EV_PLAY event will be posted when the operation is complete. The application may use the AT_TERMMASK( ) attribute to determine the exact reason for termination.

SCvox_rec( ) - *record voice data*

Synopsis

```
USHORT SCvox_rec (SC_DEVHNDL dev, SC_IOTT *iottp,
           SC_TPT *tptp, UCHAR mode);
```

Description

This function records a sequence of voice data specified by an array or list of SC_IOTT structures pointed to by *iottp*. This data is recorded until one of the terminating conditions specified in the array or list of terminating conditions pointed to by *tptp* is satisfied. These are recommended values, and may be changed. If *tptp* is NULL then previously set terminating conditions will take effect.

mode defines whether the function execution is synchronous or asynchronous. When run asynchronously, an EV_RECORD event will be posted when the operation is complete. The application may use the AT_TERMMASK( ) attribute to determine the exact reason for termination.

SCvox_playf( ) - *play a voice file*

Synopsis

```
USHORT SCvox_playf (SC_DEVHNDL dev, const char *fnamep,
        SC_TPT tptp, UCHAR mode);
```

Description

This function plays voice data stored in a file. The data in the file identified by the ASCIIZ string pointed to by *fnamep* is played until one of the terminating conditions pointed to by *tptp* is satisfied, or until EOF. If *tptp* is NULL then previously set terminating conditions will take effect.

mode defines whether the function execution is synchronous or asynchronous. When run asynchronously, an EV_PLAY event will be posted when the operation is complete. The application may use the AT_TERMMASK( ) attribute to determine the exact reason for termination.

SCvox_recf( ) - *record a voice file*

Synopsis

```
USHORT SCvox_recf (SC_DEVHNDL dev, const char *fnamep,
        SC_TPT tptp, UCHAR mode);
```

Description

This function records voice data and stores it in a file. The file identified by the ASCIIZ string pointed to by *fnamep* is recorded into until one of the terminating conditions pointed to by *tptp* is satisfied. If *tptp* is NULL then previously set terminating conditions will take effect.

mode defines whether the function execution is synchronous or asynchronous. When run asynchronously, an EV_RECORD event will be posted when the operation is complete. The application may use the AT_TERMMASK( ) attribute to determine the exact reason for termination.

Signal Computing Applications Programming Interface (SCapi) Functional Description

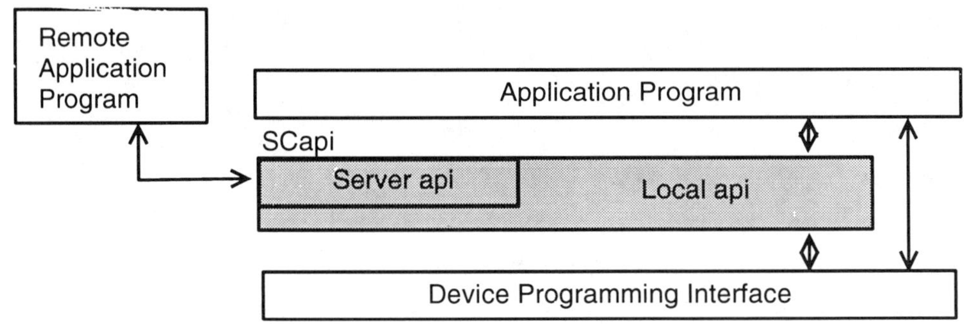

Overview

This section defines the SCapi without reference to a particular implementation. That is, it defines the semantics of a set of objects, but does not define a syntax for accessing them, or an implementation.

SCapi is an object-oriented, high level API for programming voice and telephony applications. It contains facilities to perform high-level device control (both line and local), resource allocation, first- and third-party call control, and storage management for various types of media (voice, FAX, data).

The SCapi includes a Local API and a Remote API. The local API will be documented in a later release. The section on the ServerAPI later in this chapter describes the remote implementation. The API is designed to be operating system-independent, hardware-independent, network-interface independent, and easily extensible to accommodate emerging technologies. It is an object-oriented model for ease of implementation and extensibility.

SCapi is follows an object-oriented paradigm. The rationale behind this is that object-oriented designs provide an excellent method for structuring programs . In addition, objects bind data (state) and operations together in single entity, which supports the concept of data and function inheritance, so developers can re-use what already exists.

The API defines set of object classes and creation/deletion mechanisms. This allows application developers to use the API to create objects as instances of classes and then apply operations to them. A "device/call/storage media" model accommodates traditional telephony applications as well as advanced digital telecommunications environments. In addition, the SCapi provides operating-system independence, telecommunications network independence, and system hardware independence. and is designed to support existing and future industry-standard call processing APIs.

Overview of General Architecture Features

Open specification of SC API commands and responses

This open specification allows SCSA developers to speed their applications to market because a common API enables portability. The end result is that conformance to this highest level of the SCSA architecture provides absolute application portability, and absolutely sets SCSA apart from simple bus-only architectures.

SC API provides object oriented interface and higher level of abstraction to call processing resources

These APIs are high-level libraries which abstract the application programmer from the intricacies of the underlying call processing technology. The goal in using these APIs is to make the application as device independent as possible.

Designed to interoperate with industry standard APIs

By conforming to standard APIs, interoperability is facilitated, so that applications will work with many different vendors' equipment. For example, hundreds of modems have been built to conform to the Hayes command set, which, in effect is an API. Manufacturers in call processing are also beginning to do the same thing.

SC API insulates application from underlying technology and operating system

The SC API allows developers to code applications on small development platforms, which can then be ported over to larger platforms, such as minicomputers and mainframes. This expands the marketplace opportunities for everyone who is involved in the call processing marketplace today, and allows developers to serve a greater audience of end users.

Some examples of conformance to common API structures include Rhetorex and PIKA, for example. Both Scott, VCS, and VPC conform to the same generic call processing resource API being promoted here. The idea is to broaden the scope of this initiative to include video, fax, and other technologies as well.

Since the SCapi is object-oriented, all functions are is captured in several telephony-oriented objects, which hold state and may have defined operations invoked on them. These operations cause actions to take place. All objects are instances of a certain object type, or class, which defines what the object holds, and what it is capable of doing. The API defines several of these classes, and semantics for operating on them.

Each class defines an operation as a set of signatures, which define what the operations are, what their arguments are, and what values they return. Each class may have a special operation which has the same name as the class. This is the constructor operation, which is used to define how objects of this type are brought into existence. There may be more than one of these. Note that this only defines the semantics of creation (i.e., the arguments required for creation, and the state of the created object) - it does not define the syntax or an implementation-specific way of creating objects.

Classes are defined by a formal definition, and informal descriptions of their purpose and operations. A formal definition defines the name, superclass and signatures for the class. An example of such a definition is:

```
class Dummy
  superclass: Object
  operations:
      Dummy(int)
      Dummy(string)

      int set(a: Object)
      string unset(a: Object)
end class
```

This defines a class Dummy, derived from a class Object, that has two operations: set() and unset(). Both take a single argument, of type Object. The set() operation returns an int, while the unset() operation returns a string. Dummy objects may be created from an int or a string. Type definitions can be of any defined object type, or one of the following pre-defined types:

int	a signed integer value
unsigned	an unsigned integer value
float	a floating point value
string	a character string value

Enumerated types, (sub-ranges of integer) are defined by the following construct.

```
enum <typename>
  <name>
  ...
end enum
```

The end result of API standards is that application developers can plan their solutions more effectively, and also have the flexibility to chose the best component to do the job without the worry of driver conformance. This equates to quicker development cycles.

Device Objects (Devices)

All interfaces to the telephone system, both central office lines and local telephones, are represented by a class of objects called Devices.

The API implements an abstract notion of a device, independent of the actual technology used to implement a system. This abstract entity has the properties of a Universal Port, and may have any type of resource available assigned to it on a dynamic basis. This resource is defined in terms of a particular capability required (e.g., fax, voice recognition, or MF detection capability), rather than a specific location or physical device. Thus, the API also handles resource allocation and management on a system-wide basis.

The API also allows local devices, such as telephone sets, to be manipulated. These are abstract entities as well, and may be directly attached (accessed) via a link to a local PBX switch. This transparent to the application. The telephone model is a peripheral having elements (such as buttons, a display, a hookswitch, a speaker phone) which may be manipulated under program control.

Call Objects (Calls)

The call is a central concept in telephone systems. All interactions between parties take place during a call. Accordingly, a central object of the API is a Call object, representing the occurrence of a telephone call in the system. All manipulation of a call takes place through one of these objects.

There are two possible paradigms for controlling a telephone call. Either the call is controlled by one of the participants, or it is controlled by an outside party, such as a switch or adjunct. These are termed first- and third-party call control, respectively (often abbreviated to 1P and 3P). The API supports both forms of control.

Since an end point of a call is always a device of some kind, 1P call control is done through the device object. Each device has operations to answer, reject, hold or pickup the current call appearing on the device. 3P party call control is provided by similar operations on the Call object itself.

Calls are created either by explicit creation and setup (outbound) or by an indication from the network that a call is coming in (inbound). Both of these events causes a Call object to be created, and associated with a device. Operations exist to map a call to a device, and vice versa.

Calls are directed to end points on the network, which have addresses. The API provides addressing for local devices and lines, and provides the ability to map several devices to a single address, or many address to a single device.

Media Storage

Once a call is established, an exchange of information can take place. For call processing applications, this information is often voice data. Other examples of information are image data (fax), or regular modem data. The API provides a mechanism to record, store and play back this information. This information is independent of format, and can be mixed it at will. Each element may be named, and groups of such elements may be created, and different storage strategies may be created for each group.

Event Generation and Handling

This functional area deals with the generation of asynchronous events, and provides mechanisms to pass them to an application asynchronously.

Enumerated Types

The following enumerated types are defined in the API:

ClassID

enum ClassID
OBJECT:
DEVICE:
PORTDEVICE
STATIONDEVICE
ADDRESS
MEDIA
MEDIADIRECTORY
end enum
BearerMode
enum BearerMode

```
  VOICE_3KHZ
end enum
ResourceID
enum ResourceID
  VSR_RSC
  FAX_RSC
  VR_RSC
  MF_RSC
end enum
MediaID
enum MediaID
  VOICE
  FAX
  IMAGE
  DATA
end enum
CallState
enum CallState
  IDLE
  DIALTONE
  DIALING
  PROCEEDING
  BUSY
  RINGBACK
```

Enumerated Types (continued)

```
  INTERCEPT
  DISCONNECTED
  OFFERING
  ANSWERING
  CONFERENCED
  ONHOLD
end enum
IOResult
enum IOResult
  SUCCESS
  FAILURE
end enum
```

Object Types

The following classes are defined by the API:

Object
class Object
operations:
ClassID queryType( )
void stop( )
void nullOp( )
end class

Description

Class Object is the root of the class hierarchy. It does not have a superclass. It contains basic methods need by all other objects.

Methods

ClassID queryType( )	This operation returns the type of an Object i.e. its class. This is identified by value of type ClassID
void stop( )	This operation stops any current operation on the object, if such an operation is pending and can be stopped.
void nullOp( )	This operation is the null operation. It does nothing. Its purpose is to allow object verification by ending a null message.

Device

class Device

superclass: Object

operations:

Device(name: string)
Device(desc: ResourceSet)

void configure(desc: ResourceSet)

IOResult play(m: Media, t: TermSet)
IOResult record(m Media, t: TermSet)

AddressList addresses( )

Call currentCall( )
DeviceState status( )

end class

Description

A Device represents a generic interface to the telephone system. It defines methods common to all supported devices, such as status query, configuration and operation control.

Methods

void configure (desc: ResourceSet)	This operation configures the resources required for a device. It allocates all the resources described by the argument desc and connects them to the basic device.
IOResult play (m: Media)	This operation plays the Media m on the device. The playback is terminated at the end of the message, or when any of the conditions in t are satisfied.
IOResult record (m: Media)	This operation records the media stream from the device and saves it in the Media m. Recording is terminated when any of the conditions in t are satisfied.
AddressList addresses( )	This operation returns a list of Addresses associated with a device.
Call currentCall( )	This operation returns the current call on the device.
DeviceState status( )	This function returns the current status of the device.

PortDevice

class PortDevice
superclass: Device
operations:

Call setupCall(addr: string,t: BearerMode)
void dial(s: string)

Call answer( )

```
        void reject( )
        void drop( )
        Call hold( )
        Call pickup( )

    end class
```

Description

A PortDevice object represents a general line interface to the telephone system. It may be regular POTS line, a PBX line, or some other network interface. It provides operations to setup outgoing calls, accept or reject incoming calls, and dial generalized digit strings.

Methods

Call setupCall (addr: string,t: BearerMode)	This operation allocates and places an outbound call to the address defined by the string addr.The device must be configured to accept the type of call defined by t.
void dial (A:Address)	This operation dials the digits in the string s. Valid digits are '0' through '9', '#', '*' and 'a' through 'd'.
Call answer( )	This operation answers a Call offered on a device. The Call state is changed from OFFERED to ANSWERED.
void reject( )	This operation rejects a Call offered on a device, and dropped.
void drop( )	This operation drops a Call on a device.
Call hold( )	This operation places the current call on hold. The device will be free to accept or set up more calls. The call state is changed from ANSWERED to ONHOLD.
Call pickup (call: Call)	This operation picks up a previously held call. The call state is changed from ONHOLD to ANSWERED.

StationDevice

```
    class StationDevice

      superclass: Device
      operations:
        void alert( )
        void setHook(S: HookState)
    end class
```

Description

class StationDevice represents a generic local telephone set. It has operations to control ringing and the hookswitch state.

Methods

void alert( )	This operation causes the phone to execute an alert, usually by ringing.
void setHook (S: HookState)	This operation sets the phone on or off hook.

Call

```
class Call

  superclass: Object

  operations:

      Call(BearerMode)
      Call(BearerMode,Device)

      void answer( )
      void reject( )
      void drop( )
      void hold( )
      void pickup( )

      CallStatus status( )
      Device device( )

end class
```

Description

A Call object represents a telephone call under control of the APl. All 3P manipulation of a call is done through this object. All Call objects are bound to a Device, either when they appear as an inbound call and are answered, or when they are created as an outbound call.

Methods

Call answer( )	This operation acknowledges a call, and makes it available for use by the application. The call's state is changed from OFFERED to ANSWERED.
void reject( )	This operation rejects an offered call. If other applications have registered for the call type, the call will be offered to them in turn. If not, the call is dropped.
void drop( )	This operation drops an existing call. If other applications have registered for the call type, the call will be offered to them in turn. If not, the call is dropped.
void hold( )	This operation places a call on hold, freeing the device on which it appears for another call. The Call on which this is invoked must be the current call on a Device.
void pickup( )	This operation picks up a previously held call, and makes it the current call on a Device. The Call must be in the ONHOLD state for this operation to succeed.
CallStatus status( )	This operation returns the current state of a call.
Device device( )	This operation returns the Device object associated with the Call.

Media

class Media

superclass: Object

operations:

Media(name: string)

string name( )
int length( )

end class

Description

This object stores a media stream for later playback. It may store any type of voice, image, or other data. It has a name by which it is identified, and attributes of length and type.

Methods

string name( )	This operation returns the name of the Media.
int length( )	This operation returns the length of the file.

MediaStore

```
class MediaStore

  superclass: Object

  operations:

      MediaStore(name: string)

      Media get(name: string)
      void put(m: Media)
      void update(m: Media)
      Media delete(m: Media)
      int size( )

end class
```

TermSet

```
class TermSet

  superclass: Object

  operations:

end class
```

Description

This object holds a set of terminating conditions for media play and record. These conditions are dependent on the type of media being played. This object will be documented in a later release.

Server API Functional Description

This section defines an Application Program Interface (API) for remote access to the Dialogic Call Processing Application Program Interface (SCapi). It defines the syntax of requests, certain transport issues, and a mapping to the SCapi.

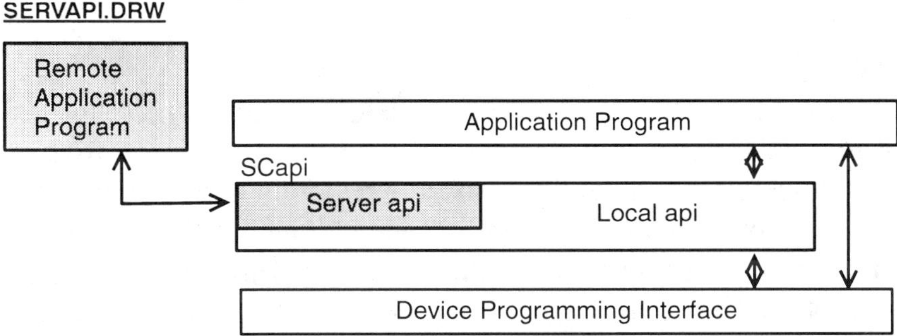

Figure **(SERVAPI.DRW)** shows the multiple levels of interaction now possible with SCSA-based APIs. Note that a portion of the SC API is divided to show the Server API, a high level programming interface that enables client or host resident application to control a call processing server. The specifications for these interfaces are detailed in the SCSA technical specification document:

- Object oriented programming interface simplifies management of call processing resources

- Insulation from underlying technology and operating system provides application portability and investment protection

- Byte stream protocol enables client or remote hosted application to control call processing resources

Overview

Server API Overview

ServerAPI enables client or host resident application control.

Native host can control an SCSA-based call processing server, allowing the application to be centralized in a mainframe or minicomputer, or distributed on a LAN. As discussed briefly before, the computing models that can be supported with the ServerAPI include *Self-Hosted, Host-Slave*, and *Client/Server*.

This is possible because the API can reside inside of a PC, which houses call processing resources. Any developer who writes application software that uses the Call Processing API can then use the resulting code to control voice processing in a native host environment via asynchronous/RS-232, or TCP/IP means. This part of SCSA has gotten a kick-start by Tandem Computers and ViCorp International, who are using the Server API to control multiple Call Processing Systems from a non-stop minicomputer.

This high-level interface can be related to the CTI, or the Computer/Telephony Interface phenomenon. The problem with most CTI implementations is that they require a major dollar investment for licensing and generic software upgrades for switches and computers. Proposed CTI standards are really not OPEN, and they entirely skirt the issue of multiple technologies. Current CTI standards also presuppose that the application resides in a host computer that drives a switch. These CTI standards do not address other core technologies such as fax, voice processing, and voice recognition.

The Remote API is an object-oriented peer-to-peer communications model that fully supports the SCapi. It is presented as a simple ASCII message protocol. The interface is based on a byte stream model, and transfers requests, replies and events to and from the SCapi using messages. A message is a stream of ASCII characters with a specific format that encodes a request, reply or event meaningful to the SCapi. These messages are independent of the method of transmission. The only assumption made is that the transport medium used is a reliable (error-free), ordered byte stream. That is, characters sent over the link are guaranteed to be delivered to the other end point, and in the same order in which they were sent. Timing constraints on the channel are not critical.

Figure **(SERV02.DRW)** displays the way the ServerAPI will evolve to include true client/server implementations. The first instantiation of the specification is limited to a host/slave model.

SERV02.DRW

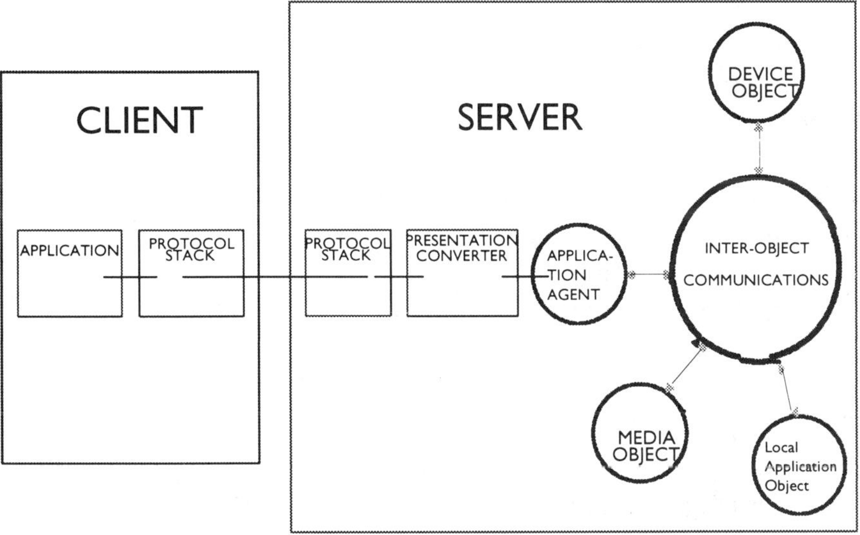

Structure

Information is transferred to and from the SCapi as requests, replies or events. A request is a message directing the SCapi to perform some action. A reply is a response to a request, indicating that the action has been done, with some particular result. An event is an asynchronous indication that something of interest has occurred in the SCapi. It is sent without prior solicitation.

Each request, reply, or event is transmitted over the link as a packet. Since the SCapi is object-oriented, each packet indicates the object involved, an operation to perform on the object, and a set of arguments. Each of these parts is termed a data element.

Packet Format

Packets are sequences of data elements separated by white space and terminated by a newline character. Data elements are described below.

Data Elements

Data elements make up the atomic parts of a packet. They represent integers, strings, object identifiers and groupings of these. They have the following representations.

Object Identifier:	Object ID
Reference Number:	Integer
Command/Event Identifier:	Integer
Command Data Items:	Any

Integers

Integers are represented as hexadecimal numbers, with an optional preceding minus sign (-) for negative numbers.

Strings

Strings are represented by a sequence of characters enclosed between double quotes (", ASCII character 0x22), such as:

"String"

Any character except a newline may be included between the quotes. To include a double quote in the string, it must be repeated. For example, the representation:

"Here is a quote - """"

represents the string Here is a quote - "". The null string is represented by a special case of two double quotes only ("").

Object Identifiers

Object Identifiers (OIDs) are represented by a sequences of the form [xxx:yyy] where xxx and yyy are (hexadecimal) integers.

Groupings

Groupings of the above elements are represented by enclosing such a group in parentheses, (and). A null data item may be represented as a pair of parentheses without anything in between: ( ). Groupings may be nested; that is, one of the grouped elements may be a group itself.

Format

Each data element in a packet has a meaning defined by its position. The format of a packet is as follows:

<objectID> <requestID> <funcID> <data1> .... <dataN>

The first data element (**<objectID>**) identifies the object involved in the message. It must be an OID. The second element (**<requestID>**) is an arbitrary integer identifying the packet. For a request packet, this number must be generated by the sender. For a reply packet, this number will be echoed (returned in the same position) for identification purposes. Any response with a request ID of 0 is an event packet, rather than a reply packet. Any request with a request ID of 0 will not generate a response.

The third element (**<funcID>**) is an integer defining either the operation to be invoked on the object, or an event identifier. Function and event codes are defined by the SCapi specification.

The remaining fields are arguments in request packets, command results in reply packets, or event data in event packets.

Formal Syntax

This section defines a formal BNF syntax for the API. Non-terminal items are enclosed in angle brackets (<>), and literal items are written in quotes (**'literal'**). Within quotes, egrep(1) regular expressions are used to define literal items.

Alternate selections are separated by a vertical bar (|). Zero or more repetitions of an item are indicated by an asterisk (**<element>***). One or more repetitions are indicated by a plus sign (**<element>+**). Curly braces group items for this purpose (**{<el1> <el2>}***). Non-printing characters are represented by C-style escapes (e.g **'\n'** is a newline character).

<packet> ::= <objectID> <reqID> <funcID> <dataItem>* '\n'
<reqID> ::= <integer>
<funcID> ::= <integer>
<dataItem> ::= <objectID> | <string> | <integer> | <group>
<objectID> ::= '[' <integer> ':' <integer> ']'
<string> ::= '"([^"\n]|"")"'
<integer> ::= <hexchar>+ | '-' <hexchar>+
<group> ::= '(' <dataItem>* ')'
<hexchar> ::= '[0-9a-fA-F]'

System Integration Description

Overview

This section describes the proposed initialization and configuration procedures for SCSA-compliant systems. The discussion includes information on the following topics:

- Initialization and configuration schemes
- Board addressing conventions
- Call fallback
- Time slot management
- Signal control model
- Conventions for managing redundant system components
- Guidelines for assessing system compliance with the SCSA standard
- System start-up sequence and the initialization functions of the Configuration Database.
- System Configuration utilities
- Board identification methodology.
- Clock recovery procedures.
- Time Slot Management Model

System Initialization Sequence

SCSA defines a start up and initialization sequence for all system hardware. For a product to be SCSA compliant, it must respond to certain messages in defined ways. The initialization sequence is divided into the following steps:

1. System Power-Up & Operating System Bootstrap
2. Locate Hardware
3. Load Hardware Bootstrap and Start Board
4. Select Boot SCbus Clock Master & Check System Configuration
5. Download Firmware and Start
6. Start Host Drivers
7. Select Runtime SCbus Clock Master and Clock Fallback
8. Perform Device-Specific Runtime Configuration

These phases are controlled by a Master Initialization Manager (MIM), using information from the Configuration Database (CDB). It is assumed that the database has been verified for internal consistency.

System Power-Up

All SCSA-compliant boards with downloadable firmware should power-up in the RESET state. SCSA boards with resident firmware in an EPROM should power up and auto-start, then wait for a message from the host or SC message bus.

Locate Hardware

Using the CDB and a vendor-supplied utility, all hardware is located in the system address space, verified for existence, then mapped into memory or IO space at the correct address. The hardware and utility should cooperate to verify that the correct hardware is present at the address provided in the configuration database. The identity of the location utility and any information it requires (program arguments) are obtained from the CDB.

Load Hardware Bootstrap and Start Board

Using the CDB and a vendor-supplied utility, a minimum hardware bootstrap is loaded on to the board. This phase will be skipped for auto-start boards with EPROMs. The utility should load the bootstrap and start the board by removing it from RESET. The bootstrap should return an indication of success or failure, and the utility should relay this, or time-out and return an error to the MIM if it does not receive it.

Select Boot SCbus Clock Master and Check System Configuration

The MIM will now select the SCbus Boot Clock Master, using information in the CDB. This device will receive a Set Clock Master command, and should then make itself clock master and verify that it has done so (see SCmessage section for a description of this). If this fails, a failure indication should be returned to the host, and the process halted.

If the board is successful in this Boot Clock Master sequence, then it should now verify system configuration by broadcasting a QUERY_DEVICES message on the SC message bus. The device should then collect all device replies, package them, and return them as part of a success indication to the MIM. The command will use a time-out parameter to determine when all boards have replied to the QUERY_DEVICE message. If the returned system configuration does not match the configuration in the CDB, the MIM will report an error, but will continue Initialization.

Download Firmware

Using the CDB and a vendor-supplied utility, the MIM will now download firmware to each board in the system. This step will be skipped for auto-start boards using EPROMs. The utility and firmware to load is defined by the CDB. Boards may be loaded by type, where several boards of the same type are loaded through a single host interface, with the firmware load being broadcast to all board using the message bus.

The CDB will determine the order in which board types are loaded, and the device to use for download. The download utility should start the download and check for a positive response from the downloaded firmware. All SCSA firmware downloads must provide this acknowledgement as their first task.

Start Host Drivers

Using a vendor-supplied utility, the MIM will start and configured the host device drivers and run-time system, and report success or failure. This now leaves the system in a state where all devices are running, and waiting for commands from the host.

Select Runtime SCbus Clock Master and Clock Fallback

The MIM will now select the Runtime Clock Master and clock fallback source, using the standard SCSA clock fallback scheme. Using the clock fallback table from the CDB, it will set the first master as the current fallback source. It will then cause a clock fallback to occur by de-selecting the boot clock master. This will invoke the clock fallback routine to select the first runtime clock master. The MIM will then select the second device in the table as the next clock master. The MIM will then record the current clock master in the CDB. See the SCmessage section for more clock fallback information.

Perform Device-Specific Runtime Configuration

Using the CDB and vendor-supplied utilities,, the MIM will now perform any runtime initialization. This may include such things as initial time slot allocations and assignments, and loading of device specific parameters (such as voice recognition vocabularies, setting system device parameters, etc). The utility to run, and its arguments, will be determined by the CDB.

SCSA System Configuration and Utilities

Within a SCSA-compliant system, many kinds of configuration information will exist. This includes such things as board types and addresses, memory locations, IRQ values, downloader and startup parameters, device names, and others. This information is required by many parts of a call-processing system. SCSA standardizes the format and storage mechanisms for this information.

Configuration Database

SCSA defines a database schema for the storage of configuration information. This schema is based on a relational model, and defines tables and fields that must be present in a SCSA compliant system. This database may be implemented in a variety of technologies, but must contain information consistent with the SCSA database schema.

Configuration API

Access to configuration information will be available via a Configuration API. This will allow applications programs and higher level APIs common access to the information.

Configuration Management

Several tools and utilities will be required to maintain and update configuration information, independent of the applications that access it. SCSA mandates only the functions required of these utilities.

Board Identification Methodology

Each SCSA-compliant board is required to have a unique ID within the system. This ID is used by the system software to locate the board within the system memory space, and to provide the board with a message bus address. The implementation of this system is product-dependent. It can be implemented using a hard-wired method, or by using programmable software registers.

System Clock Initialization and Fault Recovery

The SCSA standard allows for any one of the boards in the system to serve as the clock master at any given time (if the board is capable of generating its own clock). Therefore, it is essential that the Initialization of the system clocks, the nomination of the master clock source, and the reassignment of the role of clock master (clock fallback) be done in a standardized and coherent way.

In the following discussion; local ("intra-node") clocks and software are termed nodal; clocks and software relevant to a node other than the one under discussion are termed remote; clocks and software common to all nodes are termed system.

SCbus Nodal and System Clock Initialization

Clock Initialization depends on:

1. an agreed arbitration algorithm that allows each nodal SCbus and SCxbus to select its own master clock source.

2. the ability of an SCxbus adapter to convey message bus messages across the SCbus/SCxbus interface, even when the clocks on either side of the interface are asynchronous to each other.

Clock Initialization conforms to the following sequence:

1. A clock master is selected for each nodal SCbus and SCxbus.

2. A system master clock source is selected.

3. Once the system master clock source has been selected, the SCxbus in the master node (that is the node where the system master clock source is located) locks its clock to the master SCbus clock.

4. All nodes remote to the SCxbus master node lock their SCbus clocks to the master SCxbus clock.

During Initialization of the SCbus:

1. All electrical interfaces to the SCbus, including the interfaces to the timing signals, must be in the high impedance state. That is bus drivers must be turned off, or if combined receiver/transmitters are used, they must be set to receive.

2. The clock failure detection mechanism on each card should be disabled.

3. No board may drive the SCbus data bus. until it has been assigned a time slot number.

4. All boards capable of being an SCbus clock master will monitor the message bus, whilst executing the Initialization algorithm.

SCxbus Initialization

An SCxbus adapter is a functional component placed between the SCxbus and SCbus. The clock and data signals for theses two busses are independent but synchronous during normal operation. Conceptually, the SCxbus and the SCbus are two separated physical subnets operating using the same control principle.

During Initialization of the SCxbus:

1. All electrical interfaces to the SCxbus, including the interfaces to the timing signals, must be in the high impedance state. That is bus drivers must be turned off, or if combined receiver/transmitters are used, they must be set to receive.

2. The clock failure detection mechanism on each card should be disabled.

3. No board may drive the SCxbus data bus. until it has been assigned a time slot number.

4. All boards capable of being a SCxbus clock master will monitor the message bus, whilst executing the Initialization algorithm.

5. Until a node is instructed otherwise by the system software the local SCbus clock signals, generated by the local clock master, will be used for SCbus communication. If a remote node claims the role of SCxbus master node the nodal SCxbus adapter must inform the local system administrator software, which will then designate that SCxbus adapter as the nodal clock master.

6. In a multi-SCxbus environment a node remote from the SCxbus master node must select one, and only one, SCxbus adapter to be the master clock source for the nodal SCbus. This decision is made on command of the system software.

7. The SCxbus clock and the SCbus clocks operate independently. Therefore, there will be N+1 unrelated bus clocks for a N node system during the initialization period.

SCbus Clock Fault Recovery

In order to increase system reliability the SCSA standard provides for the dynamic reallocation of the role of clock master. This reallocation occurs either upon receipt of a command from the system software or, automatically with the detection of a clock fault. This is referred to as clock fallback.

For the case of the SCbus a clock fault is defined as:

1. Loss of one or more of SCLK, SCLKx2*, or FSYNC*.
2. Loss of Synchronization between SCLK and FSYNC*, i.e. an incorrect number of SCLKs between FSYNC* pulses.
3. CLKFAIL goes high (logical 1).

The functions of this definition are identical to that of the SCxbus, and is relevant to the equivalent SCxbus clock signals. Every SCbus board able to drive the system clocks should be capable of monitoring the status of the system clocks on the SCbus. Every SCxbus adapter should be capable of monitoring the status of the system clocks on both the SCbus and the SCxbus.

Clock fallback depends on the existence of at least one clock source in the system, other than the current clock source, which is capable of driving the system clocks. A list of possible clock sources should be kept by the system administrator software, and one source on the list, other than the current clock source, should be selected as the stand-by source. A board selected as the stand-by source is said to be armed and is capable of executing clock fallback automatically. See the SCmessage section for a detailed description of the clock fallback procedure.

The following general rules should be adhered to:

1. In the case of a clock fault on the SCbus the standby clock module should be capable of detecting a system clock fault, and initiating fallback, within 20 ms of its occurrence. In the case of a clock fault on the SCxbus the standby clock module should be capable of detecting a clock fault, and initiating fallback, within 100 ms of its occurrence.

2. Clock fallback should be accomplished within 100 ms of a clock error event.

3. During fallback the bus will be frozen, and all bus activity will cease.

Chapter 5 - Conclusions

As information technologies continue to converge and opportunities emerge for new and exciting ways of delivering information to customers, companies look for a unifying framework for integrating the diverse array of technologies at their disposal. Signal Computing System Architecture reduces the barriers to technological advances by presenting information providers with a standard way of interfacing the vast collection of communication hardware and software on the market today.

SCSA offers the following benefits:

Open architecture based on existing and future industry standards. SCSA provides compatibility with existing products as well as a growth path for new product development.
Bus with increased bandwidth, switching, and signaling capability to support the high density data transfer required by emerging technologies. Developers can reach more markets by building faster, larger systems.
Access to a wide range of signal processing algorithms and network interfaces. SCSA encourages the development and improvement of core technologies.
Well-documented interfaces throughout different layers of SCSA. Developers can create truly interoperable products at the hardware, firmware, and software level while shortening development time and getting products to market quickly.
Application programming interfaces (APIs) that increase software portability by insulating applications from the underlying technology. Developers can leverage their source code investment across different operating systems and platforms.
Flexibility to build scalable products ranging from single-node, self-hosted systems to multinode distributed systems. With minimal redesign, developers can adapt their systems to different operating environments and explore new markets.

CHAPTER 6 -
SCSA Compatibility With Other Standards

One of the philosophical underpinnings of SCSA is its ability to conform to standards that already exist, and to create in-use standards. This is bound to be controversial, since so many people in computing and communications are sold on draft international standards that are promulgated by the CCITT and by organizations such as ISO and ANSI, for example.

A De Facto standard is one which becomes accepted as a result of sheer use. A De Jure Standard is one which is promulgated by consensus *first.* The irony in De Jure standards is that they are almost always pre-empted by common use of *De Factos*, or that De Facto standards are offered up and then changed by the De Jure process.

Take MNP (Microcom Network Protocol), for example. MNP, originated as a De Facto standard protocol that provides error correction and data compression in dial-up modems. Microcom created it, and Microcom is the primary mover with regards to MNP being a standard.

According to Newton's (excellent) Telecom Dictionary, the protocol's design allows for a broad range of services to be implemented, while maintaining compatibility among modems with different levels of MNP capabilities. For example, a modem capable of MNP Class 5 and Class 7 data compression can talk to a modem that lacks MNP data compression.

Microcom says: "MNP is an error correction protocol accepted by international standards authorities (CCITT Rec. V.42). MNP offers a reliable and widely accepted method of correcting errors in transmissions over dial-up communications lines. MNP incorporates three different data compression methods, including the CCITT recommendation, V.42bis."

Since its original definition, MNP has evolved through nine classes of enhancements. Of those nine classes, the first four provide error control and are in the public domain. Classes 5 through 7 may be licensed from Microcom. Currently MNP error control (Classes 2,3 and 4) has been adopted, along with the LAPM protocol, as mandatory elements of the

Consultative Committee on International Telegraphy and Telephony (CCITT) V.42 recommendation for modem error control.

The point is that there is a hazy mix between what is De Facto and what is De Jure. Look at how long the Ethernet issue has been evolving in committees over the past ten plus years. Did Xerox sit back and wait for the ultimate specification to come out of a draft international standard before delivering value to its customer base? No, they, and thousands of others took advantage of what industry had to offer.

Such is the case with SCSA. The call processing industry already has much to offer, and SCSA is a means by which the best elements of that industry are being taken advantage of. The rest of this chapter deals with how SCSA can be made backwards compatible with PEB and MVIP-based systems.

Implementing SC Data Bus and PEB Interfaces using the SCbus Chip

The core of the data bus interface hardware is the SCbus Hardware. This section describes an implementation of the SCbus Hardware based on the Dialogic SCbus Chip (SC 2000). This implementation is capable of operation in both the SCSA and PEB environment.

The PEB Architecture - PEB DATA AND SIGNAL PATHS

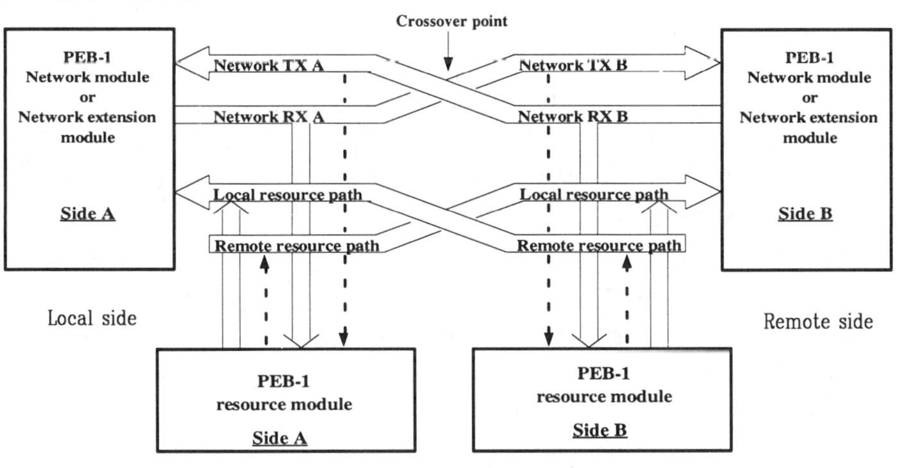

PEB data & signaling path

The PEB is a TDM bus featuring separate transmit and receive data paths. Each data path has its own set of timing and control signals. The bus bandwidth is sufficient for 24 DS0s. A DS0 time slot has a 64 kbps bandwidth, and is typically used for voice and medium speed data transmission. Figure **(SCBUS_38.DRW)** depicts the PEB timing and control signals.

SCBUS_38.DRW

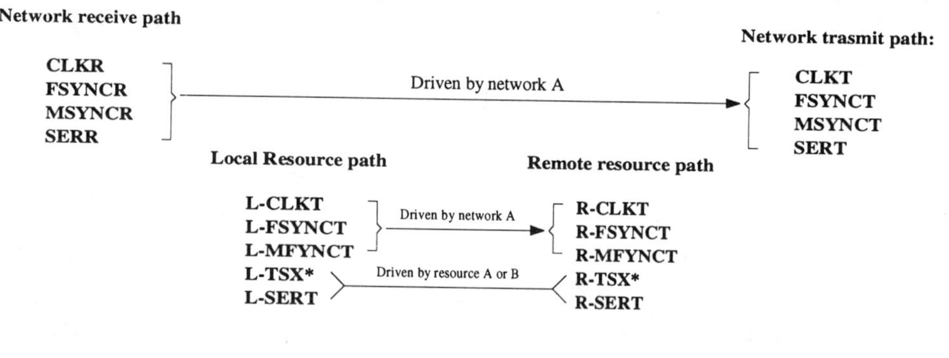

PEB bus signal

The PEB architecture is characterized by the following features:

1. All network transmit path signals (xxxT) on one side of the crossover point are physically the same as the network receive path signals (xxxR) on the opposite side of the crossover point.

2. All remote resource path signals (R-xxx) on one side of the crossover point are physically the same as the local resource path signals (L-xxx) on the opposite side of the crossover point, with the exception that L-TSX*, L-SERT, R-TSX and R-SERT are driven by the respective resource modules.

3. x-TSX* is driven by the resource module and used by a network module to read data from either SERT or L-SERT.

The SCbus Chip (SC 2000) In the PEB Environment

When the Dialogic SCbus Chip is used in the PEB environment it maintains the PEB protocol and timing requirements by propagating most timing and control signals transparently, however there are exceptions depending on the exact nature of the environment. The following section describes the 3 PEB board configurations.

Network Master Module

When the SCbus Chip is being used on a network clock master module it has the following operational characteristics:

1. Local timing and control signals are used to generate all PEB timing and control signals.
2. All timing and control signals received from the bus are propagated transparently.
3. Data is read from the PEB under the control of L-TSX* and the CPU.

NETWORK MASTER CONFIGURATION

SCBUS_39.DRW

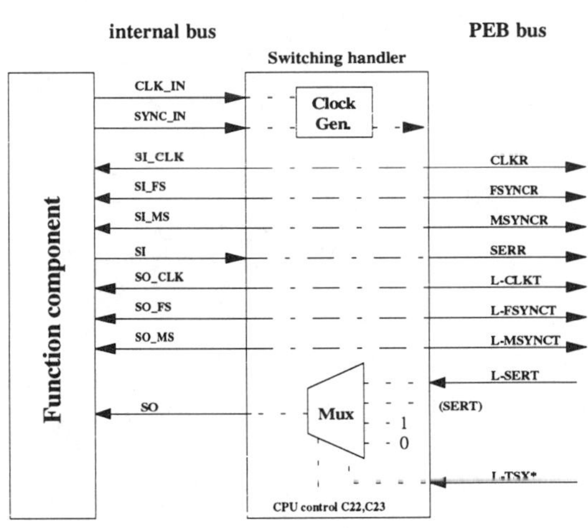

Switching handler operates in PEB network master mode

Network Slave Module

When the SCbus Chip is being used on a network clock slave module it has the following operational characteristics:

1. All timing and control signals read from the PEB are propagated transparently.

2. Data is read from the PEB under the control of L-TSX* and the CPU.

This configuration is only used when there is more than one network module in the signal processing system.

NETWORK SLAVE CONFIGURATION

SCBUS_40.DRW

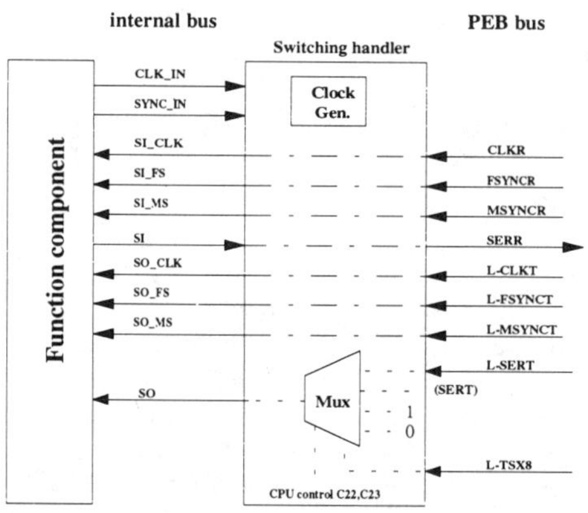

Switching handler operates in PEB network slave mode

Resource Module

When the SCbus Chip is being used in a resource module it has the following operational characteristics:

1. All timing and control signals from the PEB are propagated transparently.

2. The SCbus Chip drives both L-TSX* and L-SERT during the time slot period specified by the CPU. Outside the selected time slot period L-SERT is disabled.

RESOURCE CONFIGURATION

SCBUS_41.DRW

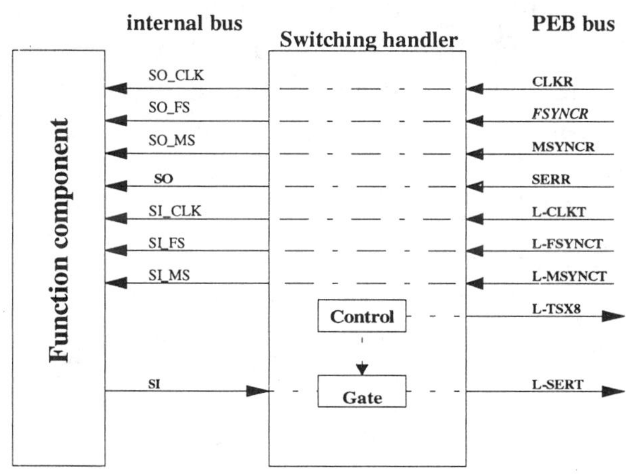

Switching handler operates in PEB resource mode

Compatibility Considerations

Although the SCbus Chip provides an effective solution for designing boards that operate in both the SCSA and PEB environment some operational constraints do exist due to differences between the two standards.

Bus Data Format

PEB Internal Format

1. There are three primary signal groups; xx_IN, SI_xx, and SO_xx. These signals may or may not be synchronous.
2. CLK_IN can be 1x, 2x, 4x, 8x, or 16x the bus clock rate. An SCbus Chip internal counter can be programmed to divide the CLK_IN to derive the desired bus speed. The bus speed can be either 1.544 Mbps or 2.048 Mbps. 2x clock is illustrated in the example.
3. SYNC_IN must be at least one clock (CLK_IN) wide. The SCbus Chip samples SYNC_IN on the falling edge of CLK_IN. The next rising edge of CLK_IN defines the frame boundary.
4. SI_FS and SO_FS overlap the first bit of each T1 frame (the F bit of the T1 frame) or the last bit of each E1 frame.
5. SI_MS and SO_MS occur once every 12 frames in the T1 environment or once ever 16 frames in the E1 environment.

PEB DATA FORMAT

SCBUS_42.DRW

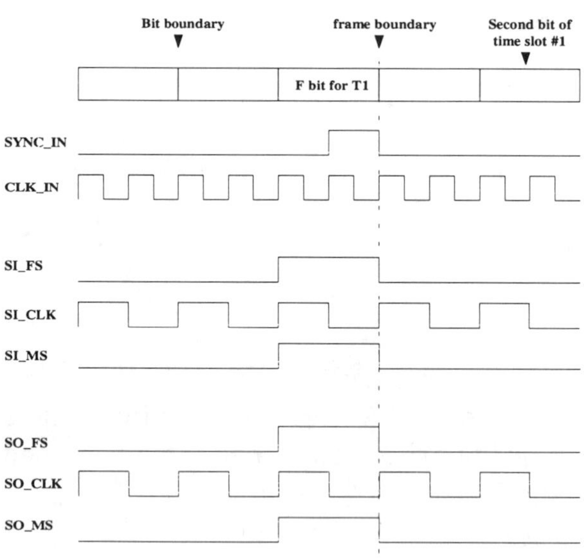

PEB internal bus format

SCSA Internal Data Format

1. There are three primary signal groups; xx_IN, SI_xx, and SO_xx. xx_IN is not necessarily be synchronous with SI_xx and SO_xx. SI_xx and SO_xx are synchronous with each ther.
2. CLK_IN can be 2x, 4x, 8x, or 16x the bus clock rate. An SCbus Chip internal counter can be programmed to divide the CLK_IN to derive the desired bus speed. The bus speed can be either 2.048 Mbps, or 4.096 Mbps. 2x clock is used in the example.
3. SYNC_IN must be at least one clock (CLK_IN) wide. The SCbus Chip samples SYNC_IN on the falling edge of CLK_IN. The next rising edge of CLK_IN defines the frame boundary.
4. SI_CLK and SO_CLK are buffered versions of 2xSCLK and SCLK respectively.
5. SI_FS is the ST-BUS frame sync signal.
6. SO_FS is the PEB frame sync signal.
7. SI_MS and SO_MS occur once every 16 frame in all applications. These signals have no "signaling significance" in the multiframe T1 or E1 environment. SI_MS adheres to the ST-BUS convention, and SO_MS adheres to the PEB convention.

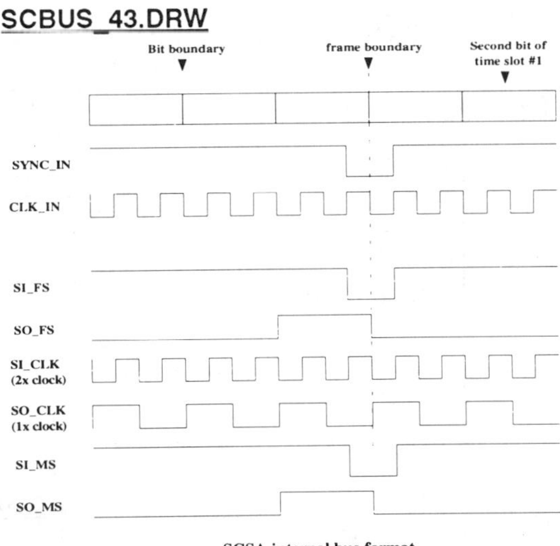

SCSA internal bus format

Operational Limitations

1. SERT is not supported. This means that a network module with SCbus Hardware based on the SCbus Chip will not support drop and insert applications.

2. R-TSX* is not supported. This means that a resource module with SCbus Hardware based on the SCbus Chip will only support a local network module.

3. SIGR, L-SIGT SIGR and L-SIGR are not supported.

4. The bus is restricted to operate at either 1.544 or 2.048 Mbps.

5. Clock fallback and the message bus cannot be implemented.

Bus Termination

To ensure interoperability and compatibility with older PEB products all PEB signals, used or unused, must be terminated per PEB system design specification. These termination values are listed in the following table.

Pin #	SCbus	Pull up by	PEB Network	Termination	PEB Resource	Termination
1	SCLKx 2*	100K			220/330	
2	GND		GND		GND	
3	SCLK	100K				220/330
4	Reserv ed	100K				220/330
5	FSYN C*	100K		470/680		470/680
6	CLKF AIL	4.7K		470/680		470/680
7	SD0	100K	L-CLKT		L-CLKT	220/330
8	GND		GND		GND	
9	SD1	100K	L-FSYNCT		L-FSYNCT	220/330
10	SD2	100K	L-MSYNCT		L-MSYNCT	220/330
11	SD3	100K	L-TSX*	470/680	L-TSX*	470/680

12	SD4	100K	L-SERT	470/680	L-SERT	470/680
13	SD5	100K		470/680		470/680
14	SD6	100K	CLKR		CLKR	220/330
15	GND		GND		GND	
16	SD7	100K	FSYNCR		FSYNCR	220/330
17	SD8	100K	MSYNCR		MSYNCR	220/330
18	SD9	100K	SERR	470/680	SERR	470/680
19	SD10	100K		470/680		470/680
20	SD11	100K		220/330		
21	GND		GND		GND	
22	SD12	100K		220/330		
23	SD13	100K		220/330		
24	SD14	100K		470/680		470/680
25	SD15	100K		470/680		470/680
26	MC	4.7K		470/680		470/680

Six 220/330 and ten 470/680 resistor networks are required. A resistor network packaged in 20 pin DIP form from Dialogic (PN 62-0013-001, see drawing below) is specifically designed for this application.

BUS TERMINATION

SCBUS_44.DRW

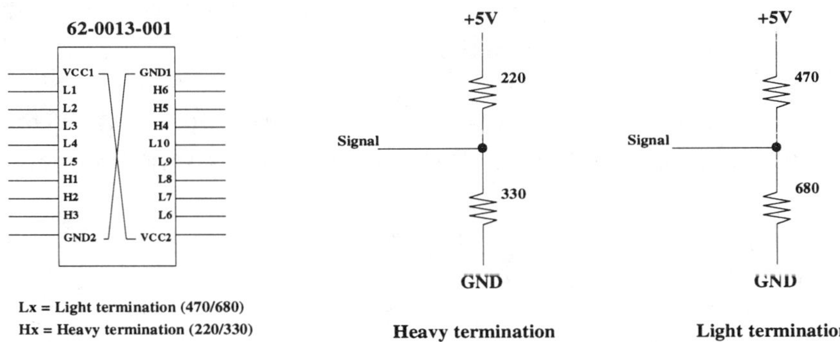

The SCbus uses a termination scheme different to that of the PEB in which all signals, except CLKFAIL and MC, must be pulled up by a 100 KW resistor. When the board is operated in a SCSA environment the PEB termination resistors must be removed.

Because termination is not always required it is recommended that the termination resistor pack is mounted on the board in a socket.

Implementing the Data Bus Interface Using the Mitel MT8980 (Mitel ST bus / MVIP)

The core of the data bus interface hardware is the SCbus Hardware. This section describes an implementation of the SCbus Hardware based around the Mitel MT8980 chip.

Data Bus Architecture and Switching Model

The distribution of data in an SCSA compliant system is based on a space-time multiplexed bus utilizing a distributed switching architecture. A fully SCSA compliant data bus interface must be capable of:

- transmitting and receiving data during any time slot period on any data stream.
- transmitting and receiving arbitrary data (that is data consisting of arbitrary sequences of 1s and 0s).
- being able to send and receive data in multi-time slot bundles (that is, a hyperchannel), whilst maintaining the sequence of the bundled time slots.

MT8980 Switching Model

The MT8980 is a 256x256 digital cross switch belonging to the Mitel ST-BUS product family. The device's switching section consists of an S/P (serial-to-parallel) converter, a 256 byte, two port, data memory, an output mux, a P/S (parallel-to-serial) converter, a control register and a pair of 256 byte connection memories.

DATA SWITCHING. The SCSA model

SCBUS_34.DRW

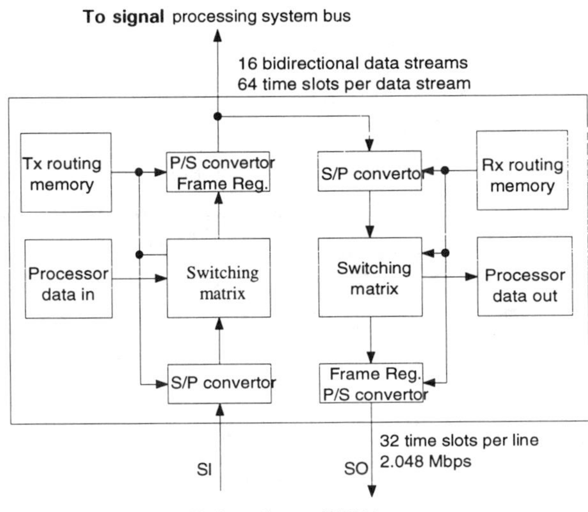

The serial input data, STi, is converted in to a parallel format (one byte per time slot) and stored sequentially in the data memory. The sequential input addresses are generated autonomously by the MT8980. The output data is simultaneously read from the data memory, converted to a serial format, and output on one of STo. The output time slot of the data is specified by the user, and depends on both the contents of the control register and the connection memory.

DATA SWITCHING. The MT8980 model

SCBUS_35.DRW

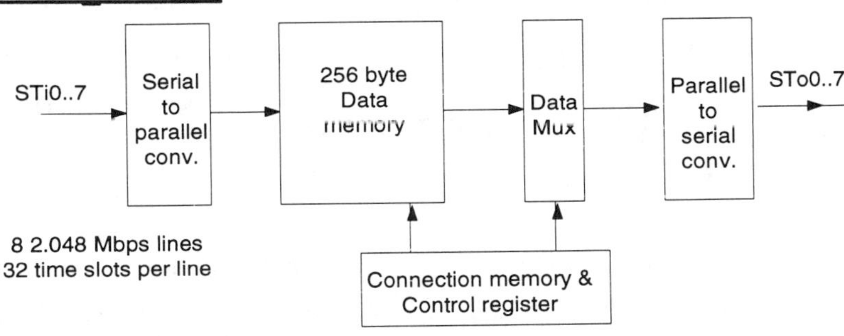

Time slot switching takes place during the generation of the output data stream. Each output time slot is associated with one particular location in connection memory. The contents of this connection memory location point to a byte of data in data memory. This is the data destined for the specified output time slot.

Interface Configuration

In order to transmit and receive data on all 16 data bus streams 4 MT8980 chips are required. The chip serial data lines (SDi and SDo) may be connected directly to the internal TDM bus and the SCbus data bus. All other signals should be buffered. All output signals must be pulled up with 100 KW resistors. A 47 or 51 W resistor in series with the outputs may be use to improve the signal quality on the bus.

For microprocessor interface and programming information please refer to Mitel application notes MSAN-119 and MSAN-123 respectively.

Compatibility Limitations

Due to limitations inherent in the design of the MT8980 not all features of the SCbus can be realized by this implementation. In particular:

- time slot sequence integrity within a hyperchannel bundle cannot be guaranteed, hence hyperchannel switching is not possible. This is due to internal delays and the use of a single data memory within the MT8980.

- bus bandwidth is limited to 512 time slots (that is an SCLK rate of 2.048 MHz).

Implementing the Data Bus & ST-BUS Interfaces Using the SCbus Chip

The core of the SC Data Bus interface is the SCbus Hardware. This chapter describes an implementation of the SCbus Hardware based on the Dialogic SCbus Chip. This implementation is capable of operation in both the SCSA and ST-BUS environment.

USING THE MT8980 to implement the SCbus Hardware

SCBUS_36.DRW

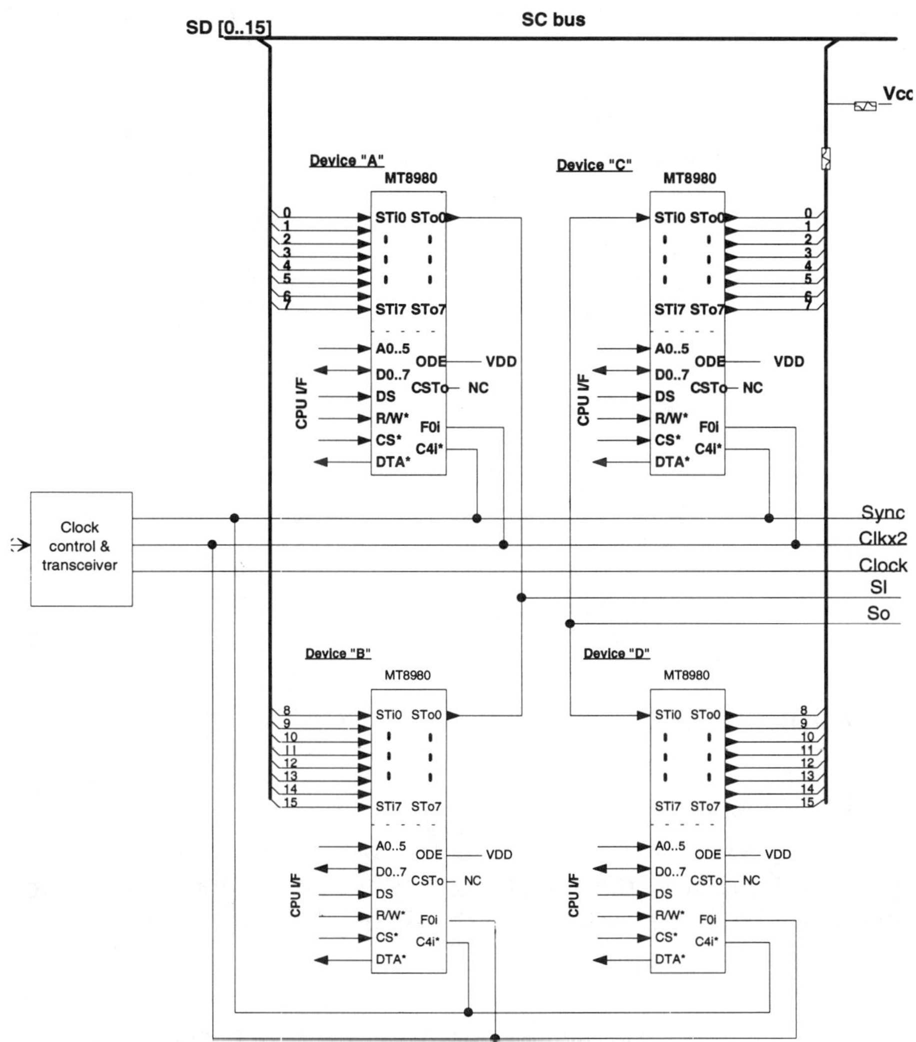

ST-BUS Architecture

The ST-BUS is a time-space division multiplexed bus featuring separate receive and transmit data paths. The bus has 16 serial data streams - 8 for

transmit data and 8 for receive data. The bus runs at the fixed rate of 4.096 Mbps. The bus bandwidth is sufficient for 512 DS0s.
The SCbus Chip In the ST-BUS Environment

When the SCbus Chip is used in the ST-BUS environment it maintains the ST-BUS protocol and timing requirements by propagating timing and control signals transparently.

THE SCBUS CHIP. In the ST-BUS environment

SCBUS_47.DRW

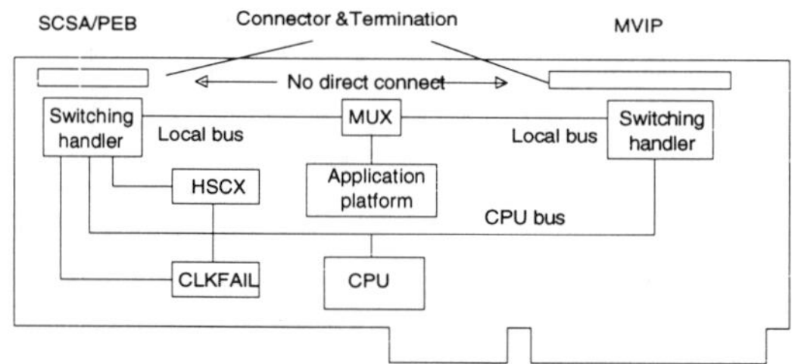

Compatibility Considerations

Bus Data Format

All timing and signal formats supported by the SCbus Chip are compatible with the ST-BUS. The ST-BUS and the SCbus have the same internal data bus format and timing requirements.

Note that the optional signals SI_MS and SO_MS occur once every 16 frames in all applications. The signals have no Îsignaling significanceÙ in the multiframe T1 or E1 environment.

The data rate of the internal bus (SO or SI) and the ST-BUS (DSo or DSi) are fixed at 2.048 Mbps.

ST-BUS INTERNAL DATA FORMAT

SCBUS_46.DRW

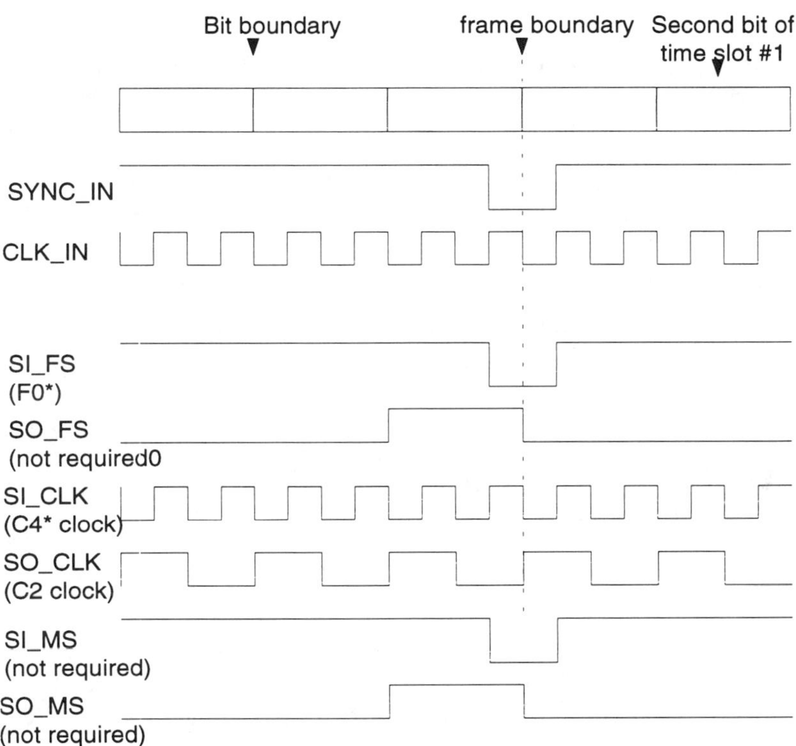

Switching Characteristics

1. Any time slot on the internal SI bus can be switched to any time slot on the external data bus.

2. Any time slot on the external data bus can be switched to any time slot on the internal SO bus.

3. Groups of time slots may be bundled to form a multi-DS0 channel.

4. In the SCbus environment any time slot may be used for transmitting or receiving data.

5. All data is subject to a one-frame delay through the SCbus Chip. D.4.3 Operational Limitations

Operational Limitations

1. $E8K_0$ is not supported. CLKFAIL is used to switch in redundant clock sources.

2. Broadcasting in the SCSA environment is implemented by transmitting on one time slot, and having all boards receive on that time slot. This contrasts with the need to transmit identical data on multiple time slots in the ST-BUS environment. The SCbus Chip does not support ST-BUS type broadcasting.

3. CLKFAIL and Message Bus operation is only supported by SCSA compatible boards, assuming the application firmware supports those functions.

Implementation

The proposed implementation provides a layer which isolates the data transport bus from the application. In addition to conventional ST-BUS switching it provides several advantages:

1. Bundled time slot switching. Current ST-BUS compatible products are unable to do this.

2. Configuration problems are avoided by the capacity to switch any time slot on the internal bus to any time slot on the external bus.

3. The distributed switching architecture available with this implementation increases system flexibility and scalability.

4. Identical internal data formats for the SCbus and ST-BUS modes enhances low-level firmware compatibility.

5. In a system with multiple nodes of mixed ST-bus and SCbus architectures, only SCSA nodes may be the system clock master.

PCB Layout

There are three major factors affecting the PCB layout: the type of connector used, the position of the connector, and the signal quality on the bus. The figure below illustrates one example of how a board capable of functioning in either the SCbus or the ST-BUS environment could be laid out.

ST-BUS/SCbus COMBINED INTERFACE BOARD
SCBUS_45.DRW

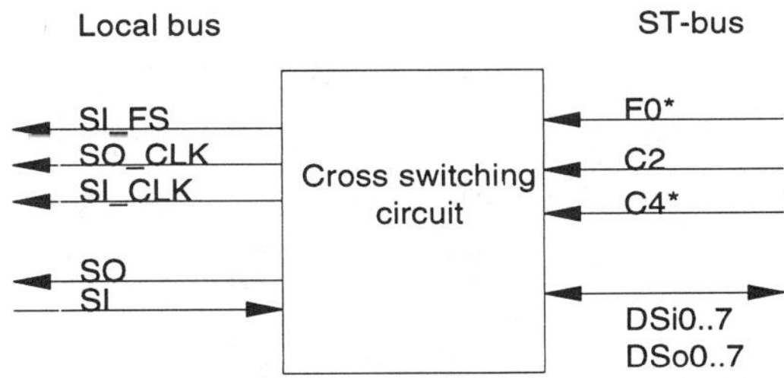

Note:

1. The need to install two SCbus Chips is avoided by the use of PLCC sockets.

2. Although both the SCbus and ST-BUS have the same internal bus formats, it is recommended that a mux is used to terminate the CPU and TDM interface for each bus. A single 16V8 PAL may be usedto achieve this.

DATA MUX

SCBUS_48.DRW

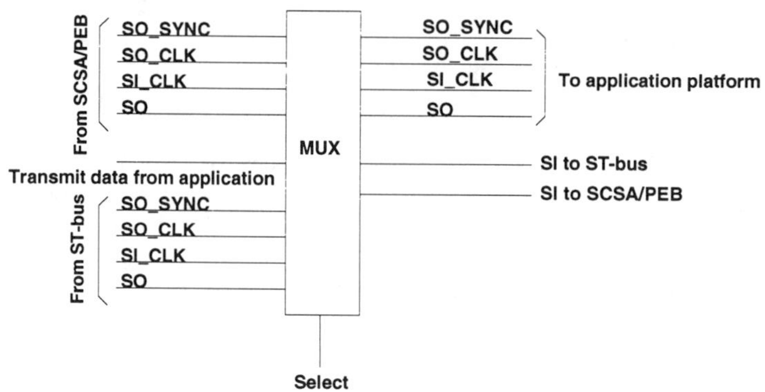

Termination

To ensure proper system operation and interoperability with other ST-BUS products each signal needs to be terminated. The input hysterisis and electrical characteristics of the SCbus Chip allow a wide variety of termination configurations.

SCbus Chip DC Characteristics

Parameter	Symbol	Min	Max	Unit
Input voltage	V_{iH}	1.77		V
HIGH - @ 4.5 V.				
Input voltage	V_{iL}		1.03	V
LOW - @ 4.5 V.				
Output High current	I_{oH}		24	mA
@ 2.4 V.				
Output Low current	I_{oL}		24	mA

Configuration

When operating in an ST-bus environment:

1. The boards base address, I/O address, interrupt, time-slot assignment, CLKFAIL enable, and Message Bus enable must be configured at system set-up time.

2. The user application is responsible for system resource allocation and assignment.

Product planning and migration

With all of the choices available to developers with SCSA, it may be confusing about just what standard to support. Many developers, such as Voice Processing Corp., Amtelco, and Acculab, for example are taking a route which allows for firmware and physical connection compatibility with both MVIP, PEB, and SCbus.

Since the SCbus is compatible with the former digital buses, its a safe bet to go with the SC 2000 chip. This provides the most flexibility in product planning and component layout and design.

Other standards explained

In addition to the PEB and MVIP busses, a variety of analog resource sharing busses are available. The most notable is the Dialogic AEB (Analog Expansion Bus). Currently, there are over 500,000 ports of voice processing installed that have the AEB connector.

The AEB is a high-impedance circuit that handles four simultaneous audio paths. It uses TTL logic state messages to handle signaling across the ribbon connector. The AEB allows for physical connection of daughter boards to the D/4x line of 4-port voice processing boards.

For example, the VR/40, Fax/40, and DTI/124 are all Dialoigc products that hook-up over the AEB. Technical specifications are available, thus enabling outsiders to make their own resource modules and/or network interface modules as with SCbus architectures. Both Rhetorex and Natural Microsystems manufacture boards that comply with the AEB standard, although these companies do not refer to the bus as AEB for competitive reasons. For example, Rhetorexs' recent announcement on 4-port voice and fax integration with GammaLink fax cards was enabled due to GammaLink's compliance with the AEB.

Other companies that support the AEB include Voice Processing Corp., Voice Control Systems, Scott Instruments, and Dianatel.

CHAPTER 7 - LANs and Client-Server Telecomputing

Perhaps one of the most exciting prospects that SCSA offers today's developer is the promise of computing and telephony confluence. Certainly, the LAN (local area network) is a key element in facilitating this merge. The LAN has also been the catalyst for a general paradigm shift in corporate computing and communications infrastructure. For example, the current computer downsizing initiatives and popularity of distributed processing in general.

Until now, voice processing and call processing were monolithic processes, which were more often than not encapsulated in some proprietary machine which had little or no intrinsic connectivity to mainstream communications infrastructures. The voice processing industry stood by the sidelines and watched as GUIs, Object Oriented Programming, and database management systems lurch forward in complexity and sophistication.

Several companies associated with voice processing are at the vanguard of a new revolution in voice processing. These are companies that are developing a new generation of integrated computer-telephony products based on the concept of the telecomputing server.

The telecomputing server integrates telephony information processing within the LAN, and more specifically, the Client-Server environment. Before we continue, let's visit the issue of Client-Server computing for a moment. A common misnomer in voice processing circles and some computing circles is that just because a platform of some kind is hooked-up to a LAN, that it instantly qualifies as a Client-Server element.

This is simply not the case. The true acid test for whether or not a process of platform is part of a Client-Server model is whether or not the processes running on the platform can be wholly separated and physically located elsewhere in the overall architecture.

For example, take a Windows-based fax program that sits inside of a desktop PC that manages personal letters and scanned images. The software in that desktop PC may interact with another machine hooked-up to the LAN that takes documents from the desktop PC and then queues them with other documents to be faxed out over common fax lines. In this example, one process is particular to an individual's PC, and the other process services many users. Both of the pieces of software are separate, yet they communicate with one another. The distinction is significant, because it is the very basis with which we design distributed processing architectures.

What is Telecomputing?

Telecomputing refers to computer-assisted telephony operations. Telephony functions for business applications includes two categories of operations:

1. Call control establishes, monitors, and controls communications links. The enterprise may make use of call-associated information such as called party and calling party identifications.

2. Media processing supports immediate transactions during a call, such as Touch Tone detection, speech synthesis and recognition, sending and receiving fax information, email-to-fax conversion, etc. Media processing may involve storage to support delayed transactions, as would be required for voice and fax messaging, for example.

When these functions are provided under computer control, this is known as telecomputing. Telecomputing has existed since the first computer-controlled telephone switches of the late 1960s. By the time the 1980s rolled around, telecomputing had expanded into a variety of application areas such as voice messaging, interactive voice response, and call centers for telemarketing.

Today, telecomputing is ready to "come into the mainstream" by truly integrating telephony information within the data processing infrastructure. The leading desktop and Client-Server architectures will play a huge role in ratifying the way telecomputing evolves over the next several years. SCSA endorsers are working very closely with the likes of Novell, Microsoft, and IBM to ensure its overall success.

LANs have become *the* way to connect PC workstations, database engines, printers, and a wide range of peripherals together. Client-Server is rapidly becoming the best way to partition large LAN-based systems into functional, easier-to-maintain segments. "Server" products respond to commands and information requests from connected workstations, hosts, and other processes via a unified command/reply protocol.

One of the many redeeming features of the Client-Server architecture is that it encourages optimized subsystems specializing in one activity or service (as opposed to monolithic mainframes that try to do everything). A good example is a commercially-available SQL server boxes that deliver fast answers to complex database queries issued on the LAN. Another benefit of Client-Server is that mission-critical applications can instantly "switch" to standby servers if the primary system fails—just for fun, try to do *that* cost-effectively on a mainframe-based system.

The open telecomputing server enables multi-vendor, scalable, and extensible applications to serve the varying needs of different groups and individuals in an enterprise. Here's the Client-Server Acid Test (four checks in right-hand column pass test):

Process	Monolithic		Multi-Process Aware	
Common Resource	Accessible By One Process		Accessible By Many Processes	
Multi-Vendor	Not Allowed		Allowed	
Scalability	By Separate Monolith		By Adding Nodes	

Call Control Standards

Powerful telecomputing applications will enhance productivity and also the level of services an enterprise may provide. Such applications will be supported by the SCSA AppServer, which is Dialogic's telecomputing server software product. Some applications that offer telecomputing server functions are approached from the multi-line client side, as with Telephone Response Technologies' Client-Server architecture.

Still other implementations will take advantage of Novell's Telephony Services API, the Microsoft Windows Telephony API, and Apple's Macintosh Telephone Manager.

What's unfortunate so far is that wholesale adoption of these initiatives is not readily apparent. There are a number of reasons for this, for example, direct desktop integration has required individual (or monolithic) desktop hardware to emulate telephones. This is unwieldy in most cases, and doesn't really fit with the idea of shared resources.

This is also particularly difficult to accomplish if the telephone instrument in question is proprietary (peculiar to a certain PBX). PBX station set interfaces vary from PBX maker to PBX maker. Add to this the complexity of matching computer application software with PBX generic loads, and we've got a real mess on our hands.

Figure DESKTOP1.DRW

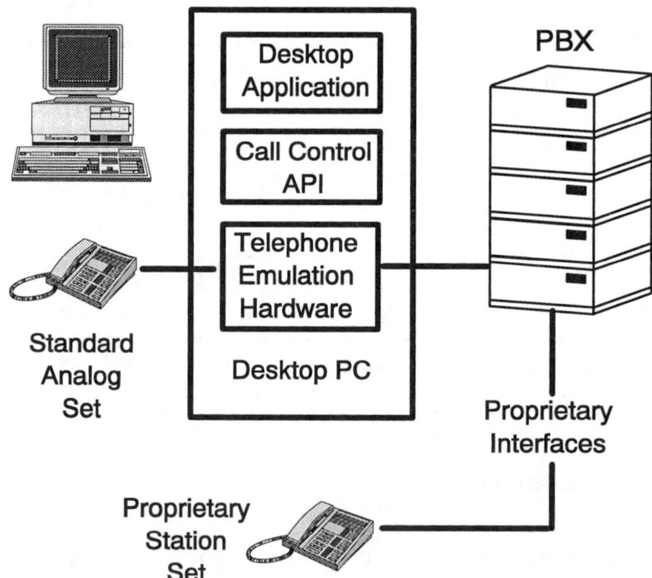

Novell, in cooperation with AT&T, is preparing a server-based solution to this dilemma. The Novell Telephony Services (server) incorporates any PBX-specific driver to provide server-based call control with an open standard interface. The Novell Telephony Services API may underlay the standard desktop telephony APIs, such as Microsoft's TAPI and Apple's Telephone Manager. This enables client applications throughout the Novell network to control the PBX through a single telephony server, instead of through special hardware for each desktop computer.

Figure DESKTOP2.DRW

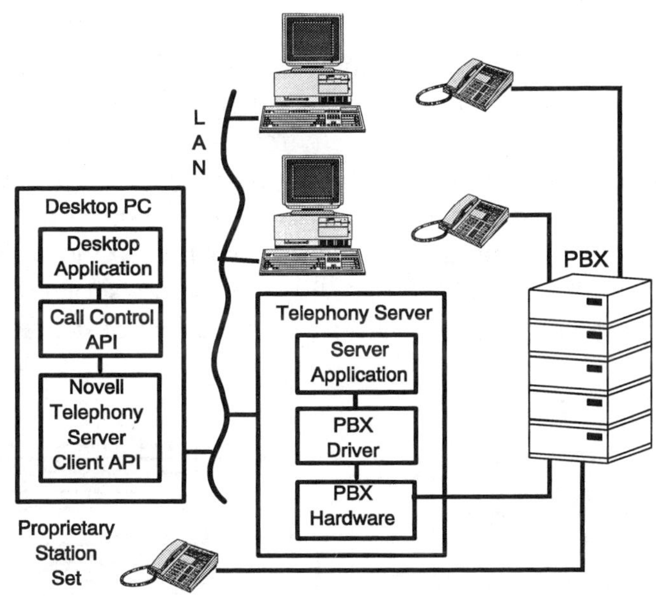

Telephony Media Processing Standards

In addition to call control, telecomputing involves the standardization of various media processing technologies. These technologies take raw communication data and transforms it into useful information that can be processed by clients in the enterprise.

SCSA helps to facilitate the use of mixed-media processes with a layered, open architecture which is built upon open standards. The SCSA approach is server-oriented, so it can be used in true Client-Server topologies as well as PBX adjunct services for classic CTI applications. SCSA incorporates call control in a manner compatible with the various call control API standards that are now evolving. The present focus of SCSA is on audio and fax media processing, and is compatible with Client-Server and desktop multimedia APIs.

The SCSA Device API is the interface to the call control and media processing applications. These Device APIs provide standard, interoperable, and vendor-independent interfaces to functions such as voice store and forward, speech recognition, telephone network signal processing, fax and data.

The SCSA Device APIs are "distributed" APIs. They are defined such that an applications written above them is readily portable between self-hosted, server-hosted, and client-hosted environments. These Device APIs are designed for compatibility with LAN messaging APIs (Microsoft MAPI, Novell MHS, etc.) and desktop call control and multimedia APIs (TAPI, MCI, etc.). A client-computer application written to the SCSA Device APIs will operate with an SCSA-based telecomputing server regardless of the operating system or hardware platform of the client or the server.

Figure SCMODEL.DRW

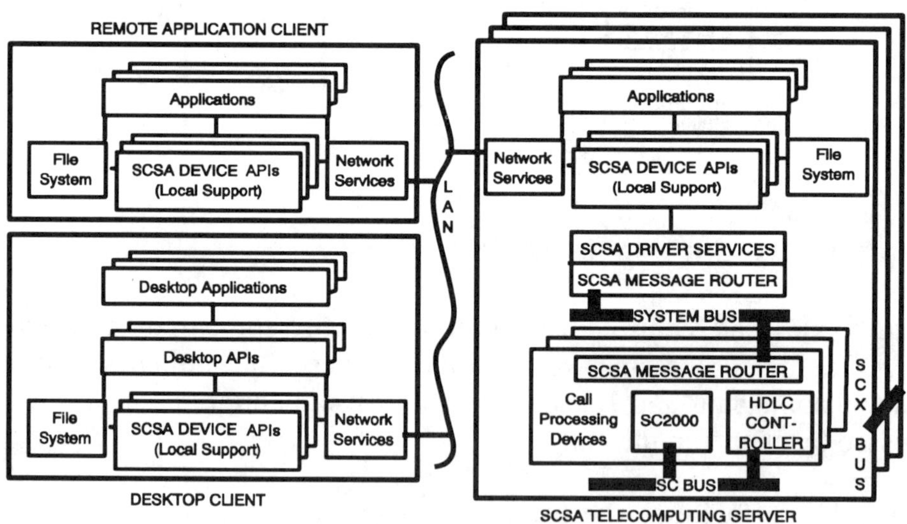

Telecomputing Server Application Examples

The Dialogic AppServer telecomputing resource provides enhanced productivity in Client-Server environments. In addition, its openness allows a variety of customized derivatives to be architected in order to meet the changing needs of today's communications and computing marketplace. The SCSA model of telecomputing can be implemented in a variety of ways. Some examples covered in this chapter will include:

- **Individual User Application Services**
- **Unified Messaging**
- **Integrated Call Center**
- **Universal Faxing On A LAN**

Individual User Application Services

Telecomputing server resources serve both the individual's telephony and audio multimedia needs without requiring special desktop-installed hardware for phone emulation or fax or audio functions. The telecomputing server can support user services such as call screening, fax management, call conferencing and transfer, voice annotation, and speech recognition.

Figure VARIABLE.DRW

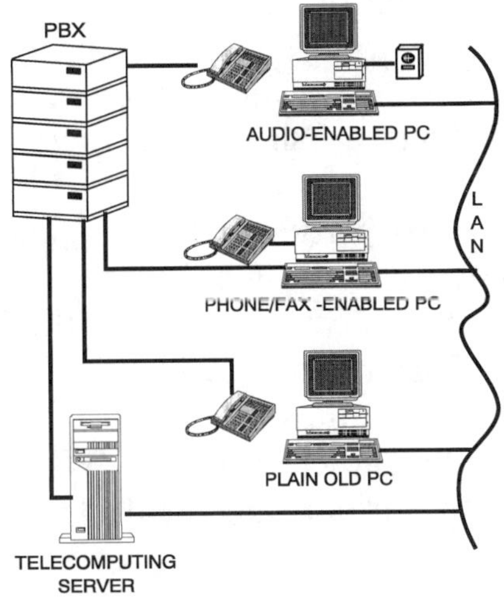

In the area of call control, the individual can manage inbound and outbound voice and fax calls from the desktop. The client APIs will provide desktop applications the ability to screen calls (identify voice vs. fax/data, determine caller ID from network signals or caller inputs, prioritize, reroute, extract the caller's database records, etc.) and to "power dial" (point and click) straight from a PC directory for voice and/or fax transmission.

Figure WINCALL.PCX

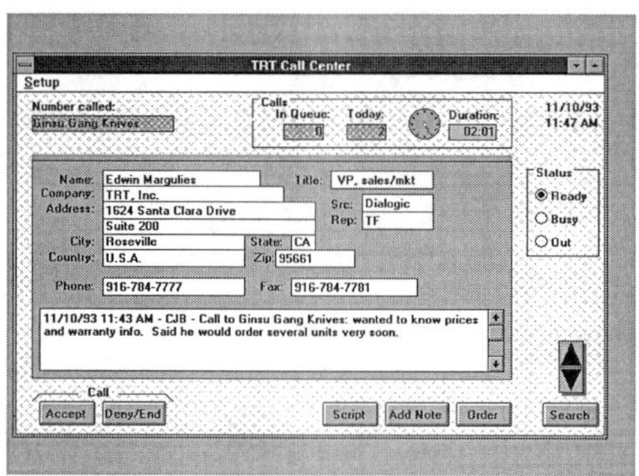

A Computer-Telephony-Integrated (CTI) interface within the server can additionally control the PBX on the same site as the client PCs. This can enable very powerful third-party control as may be desired in major call centers. The Novell Telephony Services server provides LAN-based CTI call control, for example. CTI enables authorized third-party monitoring as well as telecomputing server control of the call at all times. With CTI, the applications can conference, transfer, or disconnect an ongoing call. The CTI approach is often expensive, however, due to usury charges levied by PBX manufacturers for special out-of-band links to their switches.

Without CTI, a simple "first party call control" telecomputing server based on AppServer technology can be implemented. This approach allows for simple station set emulation in order to control the PBX. The integrated call center approach detailed later provides yet another alternative to the pure CTI approach. In this example, the applications can screen and route inbound calls as well as make outbound calls. Figure (WINCALL.PCX) shows what an agent screen looks like in an intergated environment where a desktop

Windows application has access to mainstream computing functions as well as telecomputing server functions.

Figure TELECOMP.DRW

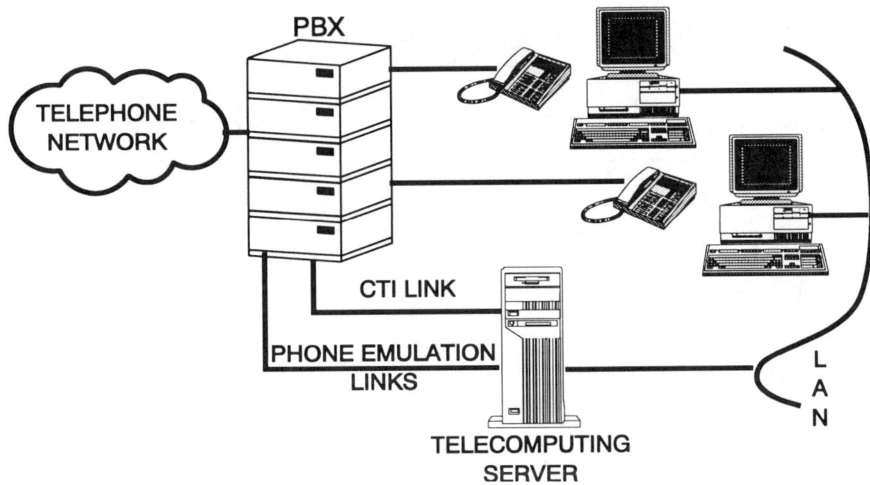

Unified Messaging

The telecomputing server enables powerful productivity enhancement when applications provide an integrated messaging interface that combines email, fax, and voice messages. Figure (UNIFIED.PCX) shows a typical unified messaging screen.

At the desktop, a single graphical user interface (GUI) enables multimedia messages to be managed as never before. "Visual Voice Mail" allows voice messages to be accessed randomly via "point and click," as opposed to serially (over the phone). The arrival time and play time for each message can be shown on the screen. The caller may be identified, and a comments section may provide the user with the ability to mark the message or topic for future reference or recall. The voice mail messages are handled jointly with email and fax messages. Capabilities include response to one medium (voice, fax, email) by using another, and the use of "folders" in which multimedia messages of various types can be bundled and stored.

Figure (UNIFIED.PCX)

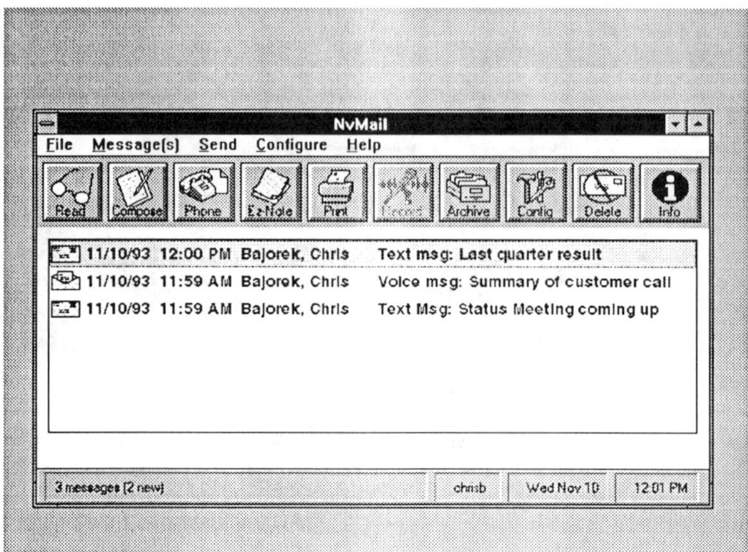

For remote access, the unified messaging application can provide media conversion. For example, a caller using a standard phone can use Touch Tone or speech recognition to scan voice mail messages, but also use these inputs to request other information to be "read out" over the phone. Such would be the case in having headers to e-mail messages alternately "spoken" with speech synthesis technology. In addition, it is also possible to have a fax converted to text by utilizing an optical character recognition (OCR) utility which converts the scanned data into ASCII text, which in turn is read out by the same speech synthesis technology as mentioned above.

Integrated Call Center

In this example, the telecomputing server eliminates the need for traditional CTI links. By splitting the application processing between the enterprise database applications and the telecomputing server itself, an interesting twist on the Client-Server paradigm can be achieved. In this case, teleservicing agents have access to a customer database via "front-end" applications that work on their desktop PC. For example, they may be accessing a Clipper database through a program like Telemagic or ACT!. This data base holds information about the callers' previous purchases, the telephone number, account number, and other identifying information which helps the agent to navigate the caller through subsequent transactions.

In many cases it is this same data base of information that is needed by the telecomputing server in order to automate the call. This approach allows the telecomputing server to use the database on the LAN as a communication I/O mechanism in order to "trade" real-time transaction data with the desktop application. A number of highly sophisticated systems of this type are being installed by systems integrators using tools such as TRT's Pro/Found database access module.

In a typical integrated call center scenario, the telecomputing server will answer a call and either collect ANI digits or prompt the caller for their customer I.D. # with the aid of a voice processing resource.

The ANI or Customer I.D. # is used as a basis for the database access module to initiate a query on the ANI database, which is located on the File Server. The search concludes that the agent associated with this customer is agent #5.

FIGURE TRTACD01.DRW

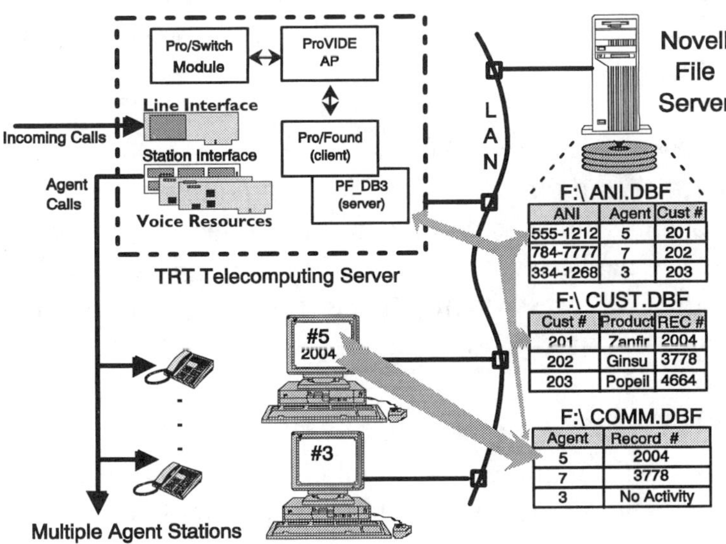

The application then uses the database access module to create an entry in a temporary communication database, by writing a record number into one of the fields associated with agent #5. This communications database is accessed by the desktop-based application once the record is released by the telecomputing server. This is enabled due to file locking and record locking conventions.

The desktop PC associated with agent #5 is attached to the LAN and is running an order entry or contact management application that uses the same database files as the database access module in the telecomputing server. The software in the desktop (agent) PC is set-up to "poll" the communications database field for a record number next to its agent number. If the field has data in it, the application "pulls-up" that data through whatever graphical or character-based method is provided by the application. This could be Telemagic, ACT!, or a FoxPro for Windows database program, for example.

The desktop PC associated with agent #5 is then available to the agent to make changes to the database for that customer as appropriate. Once the screen is filled with the customer record information, the telecomputing server will route the caller over the station interface to the telephone associated with agent #5. This can be done as a result of an automatic routine in the agent's application in his or her PC that writes new information into the communications database, or manually by the agent by striking a key, etc.

Universal Faxing On A LAN

Another viable scheme for LAN-based telecomputing is the multi-port client-as-host approach. This is manifest in Telephone Response Technologies (TRT) popular Client-Server architecture. A new product called *I/Fax Plus* exemplifies the way in which shared resources and distributed client computing can work together.

Modern (fax) problems

Corporate fax traffic is increasing dramatically and with it come a host of new problems. The single fax machine in a central location (or one fax machine in every corporate department) is becoming difficult to justify.

Why? Because increased fax traffic means longer lines at the fax machine and lost time. Installing single-line fax cards at employee's workstations helps but usually the added expense and support burden of the extra phone lines and fax cards quickly becomes unacceptable.

Another need for fax arises because many corporations are now delivering faxed information ("demand publishing") to their clients via interactive voice response access—anyone with a Touch-tone phone can get access to data sheets, pricing, order status, etc. Some corporations routinely distribute periodic information via mass "fax broadcasts." For others, the timeliness with which incoming faxes are delivered can be critical (e.g., incoming orders). These seemingly dissimilar needs for sending and receiving fax information can be centralized so that employee productivity is improved, money is saved, and the management and support efforts are optimized.

Enter the fax server. Install one on your LAN and employees can send and receive faxes from their PCs without leaving their desk. With a fax server, interactive voice response (IVR) systems with special capabilities can send faxes. Lower costs are realized since the fax server resources (i.e., fax cards and fax phone lines) are *shared* across all users. Centralized control and administrative functions exist, and it is easier to add fax lines to meet increased demands later.

Before/After—A typical corporation

Let's look at how a fax server can save time and money. The traditional way was to have several fax machines located throughout the company where employees took their printed materials to be faxed. As well, the company's IVR system had its own set of fax cards used to deliver important product information to callers 24 hours a day. By using a fax single fax server, *direct* desktop access by all employees is enabled. In addition, IVR access to those same fax resources means fewer phone lines and better fax resource management, reporting, and control.

A new approach: I/F*ax Plus* Architecture

The *I/Fax Plus* product seamlessly combines "fax from the desktop" needs with "fax on demand" services. Functionally, *I/Fax Plus* is divided into two parts: the **fax server** and the *I/Fax Plus* **add-on module**.

Figure IFAXPLUS.DRW

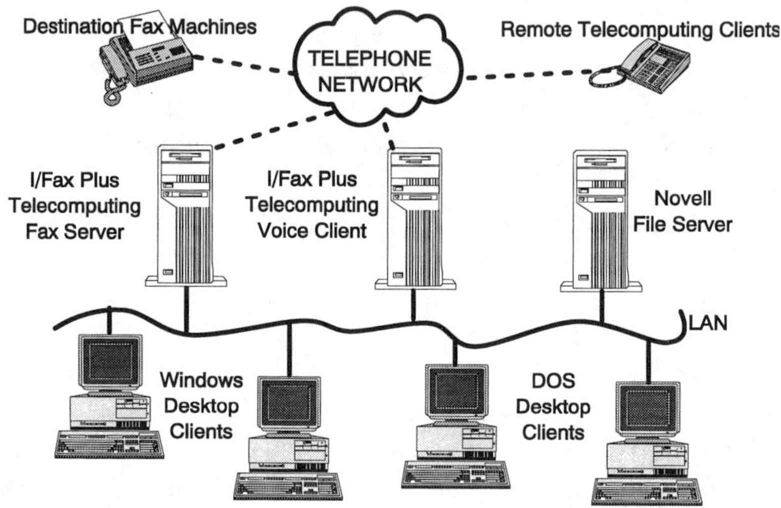

The **fax server** software loads on a dedicated computer on a Novell LAN and it supports multiple fax lines. Supported fax cards include INTEL, Hayes, GammaLink, BrookTrout, Dialogic, and many others. Fax sending and receiving from desktops is accomplished by using Microsoft Windows "front-end" programs like the feature-rich FaxWorks/Pro from SofNet. More power: any DOS or Windows program that can print to a selected Novell print queue can send a fax (e.g., Word for Windows, Excel, WordPerfect, etc.). Graphics, raw text, and Postscript files can also be faxed. Cover page support and fax "phone book" management are standard.

The second component, the *I/Fax Plus* **add-on module**, gives fax capabilities to the TRT ProVIDE product making it possible for Touch-tone callers to send faxes via the *I/Fax Plus* fax server. This approach means *no* fax cards are duplicated in the IVR system, yielding more free card slots and less processing overhead. Advanced features like full fax phone number validation, fax priorities, and future fax delivery are included.

CHAPTER 8 - Telephony and SCSA

Network Interfaces, Switching, & Resource Modules

As discussed in chapter one, voice and call processing represent the intersection between computation and telephony. In that chapter, we discussed the relevance of PBXs, communications in general, and protocol converters. In this chapter we'll take a closer look at the fundamental elements in voice processing, and how they are controlled or modeled from an application view.

Figure **(BLOCK01.DRW)** provides an illustration of the separation between the line interface, shared resource bus, and signal processing on a generic voice processing board.

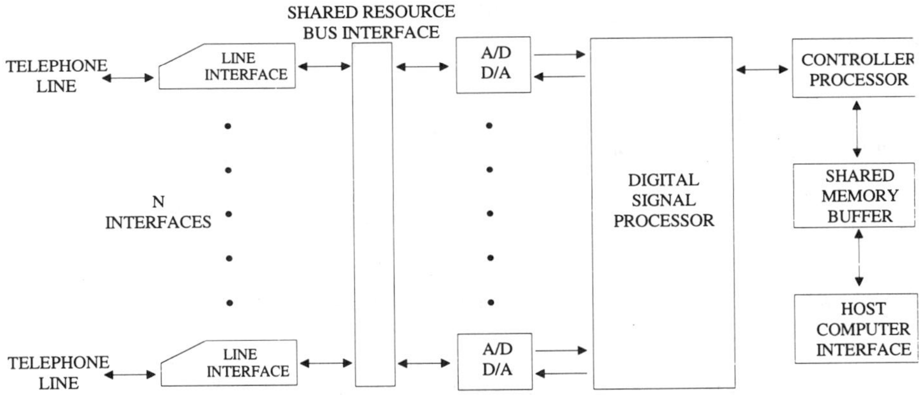

The line interface technology used on boards like this one is not at all dissimilar to the ones used in PBXs and key systems. In fact, the resource sharing bus that connects voice processing boards can be likened to a switch back plane. Until SCSA came along, the distinction between a switch's back plane and what you could do in a voice processing platform was always the time slot handling capability of the bus. This is no longer a huge distinction, since the SCbus easily eclipses over 50% of all of the key systems and PBXs on the market today in terms of bandwidth.

The diagram also point out that the digital Signal Processor, line interface, and analog to digital conversions are controlled by a control processor. Memory is also pictured, as is external interfacing devices. Let's take a look at what elements go into a typical PBX or key system now and compare them to their equivalent in the SCSA architecture:

SCSA and Generic Switching - A Comparison

SCSA Architecture Equivalent	Common PBX/Key System Component
Host Computer	Control Processor
Host hard disk	Program Store (disk/ROM)
Host Random Access Memory	Page Memory
Serial Ports / LAN	Peripheral I/O
SCbus / SCxbus	Switch Back plane
Trunk Lines / Station Lines	Trunk Lines / Station Lines
DTMF Detection	DTMF Detection
Voice Processing	Not applicable
Fax Processing	Not Applicable
Voice Recognition	Not Applicable
Speech Synthesis	Not Applicable

As you can see, with the addition of PC-based components which drive screen-based telephones, there's not a lot a switch can do that a call processing built on SCSA can't do. As a matter of fact, there's a lot switches can't do that an SCSA-based call processing system can do.

The concept of the Business Communications Controller

Recall the diagram in the second chapter which illustrated the integrated voice response and switching system in a 48x24 agent configuration. This is an example of how call processing systems can provide multiple functions to the user. For example, network line access, common switching services, and station (telephone) support.

Imagine adding to these capabilities a host of other core technologies such as facsimile, voice processing, and speech synthesis. This is what is called the Business Communications Controller. Figure **(VPBLOWUP.DRW)** depicts the diversity of technologies that can be merged by using SCSA-based components and busses.

VPBLOWUP.DRW

CROSS SECTION OF A MULTI-TECHNOLOGY VOICE PROCESSING SYSTEM

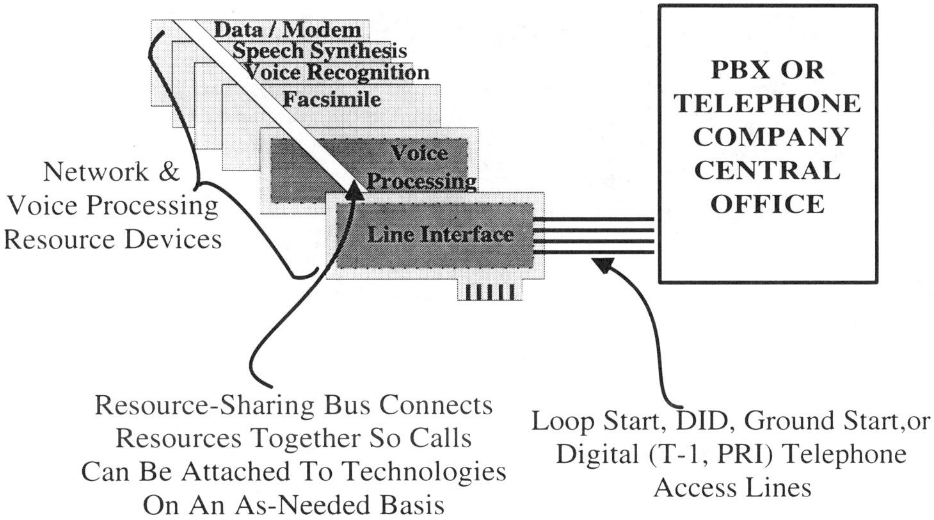

Data / Modem
Speech Synthesis
Voice Recognition
Facsimile
Voice Processing
Line Interface

PBX OR TELEPHONE COMPANY CENTRAL OFFICE

Network & Voice Processing Resource Devices

Resource-Sharing Bus Connects Resources Together So Calls Can Be Attached To Technologies On An As-Needed Basis

Loop Start, DID, Ground Start,or Digital (T-1, PRI) Telephone Access Lines

Figures **(BOARD1.DRW)** and **(BOARD2.DRW)** show how the resource cards inside a Business Communications Controller are put together. In the former, both the line interface technology and the core voice processing or other processing technology are housed on the same card. In the later case, the line interface and core technologies are on separate cards.

In both cases, the cards have a resource sharing bus connector which allows a mezzanine cable to connect all of the boards together. This bus carries the signals for SCbus, PEB, MVIP, or AEB, for example. In effect, the resource sharing bus is an extension of the telephone network, because the switching of resources in and out of use can be achieved by host control and signaling on the bus.

In some cases, line interface boards "surround" voice processing boards for a drop and insert configuration. This means that a telephone call can enter a call processing platform from the network, and then get some kind of treatment in the platform before being switched elsewhere.

BOARD1.DRW

MAJOR ELEMENTS FOUND ON A DUAL-FUNCTION VOICE PROCESSING RESOURCE CARD

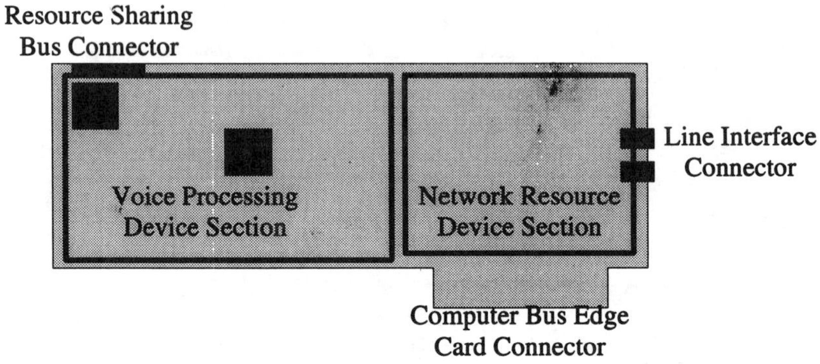

Resource Sharing
Bus Connector

Line Interface
Connector

Voice Processing
Device Section

Network Resource
Device Section

Computer Bus Edge
Card Connector

BOARD2.DRW

MAJOR ELEMENTS FOUND ON SEPARATE VOICE PROCESSING AND NETWORK RESOURCE BOARDS USING A SHARED RESOURCE BUS

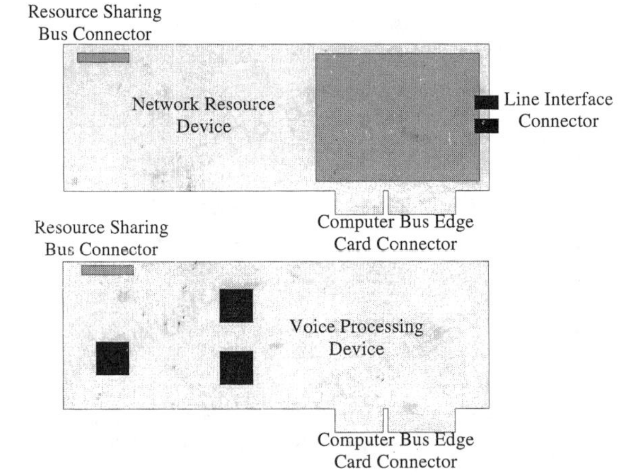

Resource Sharing
Bus Connector

Network Resource
Device

Line Interface
Connector

Computer Bus Edge
Card Connector

Resource Sharing
Bus Connector

Voice Processing
Device

Computer Bus Edge
Card Connector

Figure (D&IO.DRW) depicts the common way of achieving this in a voice processing platform. This makes possible a variety of sophisticated applications such as the ones that were discussed in chapter four.

The type of signaling on one side of the drop and insert configuration can be different from the other side, much the same way a protocol converter operates. For example, a PRI (Primary Rate ISDN) circuit could be connected on the front, and a regular T-1 circuit could be connected on the back. This could provide for network connections to premium services such as AT&T Info2, but with a telephone system that is not directly compatible with PRI.

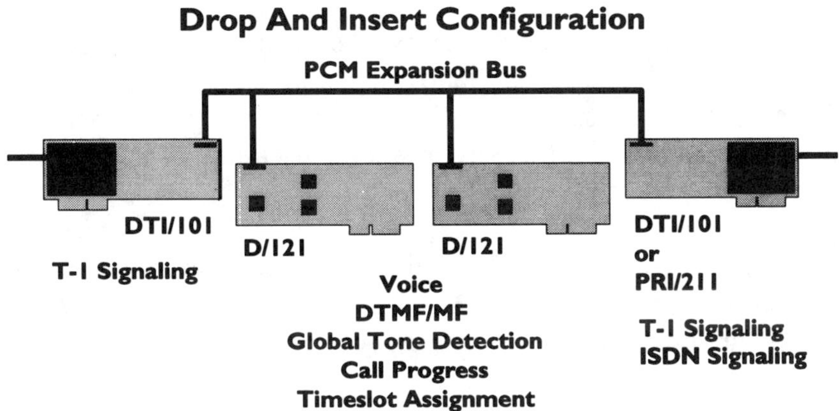

The multi-port Telephony-based model

This section describes the SCSA Application Model. This model is based on a common switched system paradigm. SCSA provides a software architecture and model for building call processing applications on top of this paradigm.. This model is the basis for both the SCdpi and SCapi specifications.

The model is designed to provide a framework and set of rules that make the task of writing a call processing application easier and make the application more portable. This framework is based on an analysis of some of the common tasks done when writing call processing applications, and contains services and facilities that provide these common tasks in a useful and flexible manner.

Definition of Terms

resource: Any hardware or software service used to control and process calls. Examples are voice store and forward, fax send and receive, text to speech conversion, voice recognition, etc. A **network resource** is any resource that interfaces with the telephone network. These may be divided into analog (loop start, ground start, etc.) and digital (T1, E1, ISDN) network resources.

connection: a TDM data path between two resources. It connects the inputs and outputs of the two resources, and may be unidirectional (simplex) if either of the resources has only an input or an output; otherwise it is bidirectional (dual simplex). It usually has a bandwidth that is a multiple of a DS0 channel.

resource group: A dynamically configurable set of resources that are used within an application to service a call. Resources in a group are allocated to that group, and are not available for use elsewhere.

universal device: a resource group with a set of connections defined on it such that the group as a whole can be treated as a single entity for the purposes of processing a call.

The Application Model

The SCSA application model is a simple one based on the idea of forming resources into groups, and defining connections between these resources. Groups collect related resources (almost always on a per-call basis), define their interconnection, and allow these resources to be treated as a single entity and operated on as a unit. They also define a unit for handling events from group members.

For example, an incoming call might need voice recognition services and a send-fax capability. The application model permits the dynamic creation of a logical grouping of resources containing the technologies necessary for handling the call. Applications use call processing resources by creating such a group of resources and defining the necessary connections, on an as-needed basis for each call.

An application therefore processes a call by creating a group, adding resources to the group as required, defining some standard data path connections between the resources, and performing call processing operations on the group as if it were a single entity. At any time, it may add or remove resources. It may also temporarily override data path connections, and form new connections between members, both intra- and inter-group, to implement such features as transfers and conference calls.

This simple model has a number of advantages. Because an application can treat grouped resources as a single entity, and this entity may have any resource available attached to it, the application may view such a group as a universal device. Groups and connections can thus be seen as a mechanism for creating virtual universal devices: single, integrated entities that have all the resources required to process a call.

Since these universal devices can be dynamically configured with resources as and when they are required, resources may be shared, and used much more efficiently. Since resources are only attached to a call as needed, a smaller set of resources may be used to service a given number of calls than if resources had to be pre-allocated. If required, a pre-allocation strategy can also be used.

As part of the model, a connection hides much of the detail of the underlying switch matrix, and relieves the application of having to track and maintain complex time slot assignments and relationships. Instead it may work with a set of default connections, which it may override with complete control as and when needed.

These two features -- the resource group and the connection model -- allow much greater simplicity in applications, relieving them of the burden of low-level administration tasks, and allowing them to concentrate exclusively on solving the problem at hand.

Creating and Destroying Resource Groups

Resource groups are created by *claiming* an initial resource. This creates the group, and adds the first member. When a resource is claimed by creating a group, it is made unavailable for further allocation.

Once a group is formed, other resources are added by *attaching* them to the group. This makes them unavailable for further use until they are detached or the group is released. It is illegal to detach the resource that was first claimed to create the group; this resource can be freed only by destroying the group as a whole.

When a group is destroyed (for example, at the termination of a call), all attached resources are detached, the initial resource is freed, and the group ceases to exist.

Groups and resources are identified by *handles* returned from the claim and attach operations. Both groups and individual members may be operated on by resource specific operations (see below).

Defining Connections Between Members

Attaching a resource to a group can define a data path, or connection, between it and another member of the group. Depending on the nature of the two resources, either a single or dual simplex connection between the two will be maintained for the duration of the group.

Connections between members must follow a many-to-one relationship, and thus can only form a tree structure.

While the underlying technology might not support the simultaneous connection of several outputs to a single input (as is the case with SCbus), the switching model will take care of any underlying switching operations needed to share the single input between the several connected outputs. This effectively time-shares the input, and gives the illusion of several outputs permanently connected to the single input.

Attach connections will be maintained throughout the life of the group, but may be temporarily overridden using a separate *connect* operation. This operation sets up a temporary connection between resources. A temporary connection exists until overridden by another connect, or removed by a corresponding *unconnect* operation.

The connect operation does not remove any connections formed by attaching resources; it simply overrides them. When these temporary connections are removed, the original, permanent connections are restored.

It is not necessary to define a connection in order to attach a resource. It is quite possible for the application to request that no connection be made, and manage the connections explicitly, resource by resource, using connect operations.

These features give applications flexibility in maintaining connections between resources. Fixed sets of connections may be set up once, at attach time, and selectively overridden as required. For more complex cases, resources may be attached without connections, and all connections managed manually using explicit connect operations.

Operating on Resource Groups

Once a resource group is created, an application may operate on it as a single entity. At the SCdpi level, it may apply any technology-specific command to a group, and within a set of defined circumstances, have the interface infer the group member that the operation should be applied to. In this way, the application does not have to track each resource separately.

At the SCapi level, the model is even simpler: operations are presented in an application-specific, rather than a technology-specific, context, with the underlying implementation doing even more work at deciding just what operations should be invoked, and what they should be invoked on.

Using Resource Groups in an Application

The resource group and connection mechanisms have been designed such that the majority of applications will use them in a simple and pre-defined manner. This section describes this usage. These mechanisms have been designed to be as flexible as possible, and there are many other ways of using them.

FORMING A RESOURCE GROUP

APPMOD1.DRW

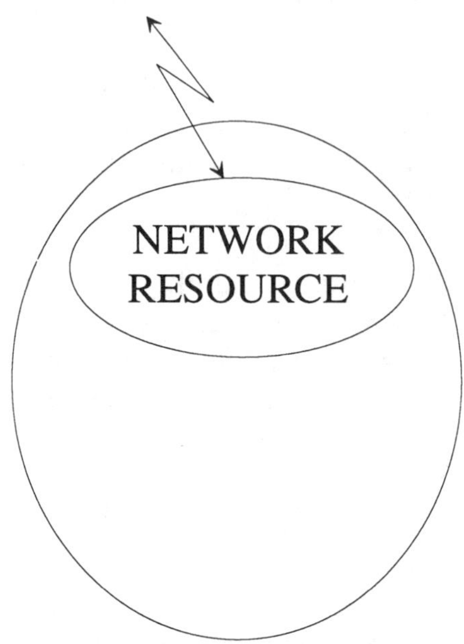

FORMING A RESOURCE GROUP

In the most usual case, a group is formed by claiming (as the initial member) the network resource on which the call will be placed or accepted (see Figure **(APPMOD1.DRW)**). To this are attached any needed resources, such as Voice Store and Forward, which are all connected to the network resource member. Note how the correct unidirectional or bidirectional connections are made, depending on the type of resource connected.

Once created, the whole group can be treated as a single object, and have commands applied to it without having to specify which member resource should execute which command. The rules defined by the model ensure that commands are directed to the correct resource, and the connections defined within the group ensure that this resource is connected to the call (via the network resource) when it is activated.

ATTACHING RESOURCES TO A GROUP

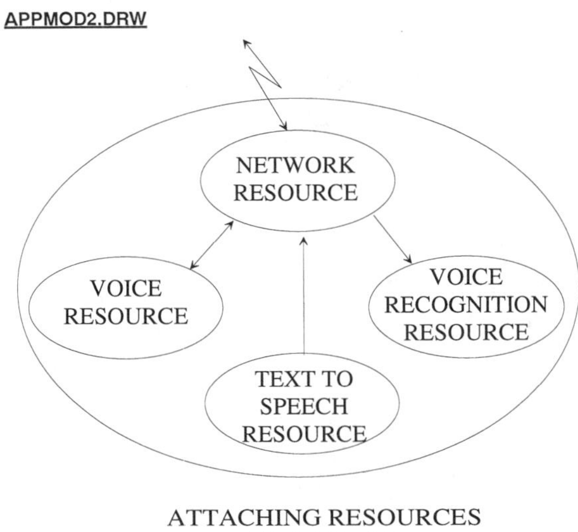

ATTACHING RESOURCES
TO A GROUP

Resources do not have to be pre-allocated by being attached when the group is created; they can be added (or removed) dynamically as a call progresses.

When processing has finished, the application may either release the group to free all used resources, and create another group for the next call, or it may detach any resources that it deems necessary, and reuse the group for the next call.

More complex configurations are also possible. An example of this is a drop and insert configuration, such as between a network interface and a local PBX. In this case, two groups would be formed: one for the network (upstream), and the other for the PBX (downstream). These two groups separately perform the required network and PBX processing until the call is passed through. This is pictured in fiigure (APPM003.DRW).

At this point, the two groups must be joined to pass the call through. This is done using an explicit, inter-group connection. By applying a connect operation between the two network resources, a temporary data path is formed, passing the call through. This is shown in Figure **(APPMOD4.DRW)**.

PREPARING FOR DROP AND INSERT

APPMOD3.DRW

PREPARING FOR DROP AND INSERT

TO NETWORK

TO PBX

NETWORK RESOURCE

NETWORK RESOURCE

VOICE RESOURCE

VOICE RESOURCE

NETWORK GROUP

PBX GROUP

Once the inter-group connection is established, the attached resources do not play any part in the process, since their connections have been overridden. They may be detached if not required further. The temporary connection stays in place until explicitly removed by an unconnect operation, when the pass-through path is broken. At this point, the groups are again separate, as shown in Figure **(APPMOD3.DRW)**, and may be released or reused, as for the previous example.

APPMOD4.DRW

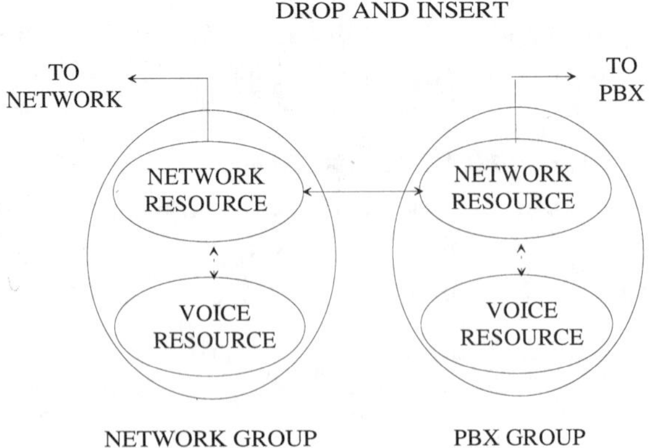

CHAPTER 9 - SCSA in Native Host Environments

The Demystification Of Telephony

Telephony is undergoing a general "opening," much like the computer industry has experienced over the past ten years. This is being forced by buyers, systems integrators, and computer manufacturers who wish to apply more application-level control over switching and line access functions.

This has not been easy, given the historical attitude of the switch manufacturers: *"Switching is the domain of the very few. It is a black art that only we understand."* The voice and call processing industry has done much to dispel this premise, and there are a number of fairly young CTI (Computer/Telephony Integration) initiatives underway that will change the way we look at switching in the long term. CTI provides for one or many openings in a telephone system that lets you link a computer to that phone system. This link allows commands from the computer to drive the phone system to answer calls, put calls on hold, and transfer calls. The term is also called OAI (Open Application Interface). The term OAI was first used by PBX makers NEC and InteCom and now the term has become somewhat generic. Until CTI, all phone systems used proprietary and closed architectures. The chart below compares the "openness" of Telecommunications equipment such as key systems and PBXs against Computer equipment and SCSA-compliant signal computing gear:

Discipline	**Computer Equipment**	**SCSA-based Equipment**	**Telecom Equipment**
Applications Software	YES	YES	NO
Operating System	YES	YES	NO
Networking	YES	YES	YES
Common User Interfaces	YES	YES	NO
Printed Circuit Cards	YES	YES	NO
Processors	YES	YES	Partially
Semiconductors	YES	YES	YES
Peripheral Devices (Disks, Control Panels, etc.)	YES	YES	NO

Virtually every notable PBX or Key System manufacturer is evolving towards open application interfaces of their own. According to Probe Research, there are really two separate "markets" for CTI:

First, there's transaction-driven applications. These are applications that support an actual business transaction such as credit collections, locator services, inbound telephone order taking, and customer service. It is typical for these kind of applications to require computer database access. The database can be on a local or remote host, and access to the data is almost always time-sensitive.

Second, there's horizontal Office Automation applications. These are applications that support business functions across organizational groups or industry verticals in inter- or intra-department business settings. Examples include productivity-enhancing capabilities such as e-mail, voice mail, and corporate telephone directories.

In addition to the switch manufacturers, virtually every major computer manufacturer has joined the CTI bandwagon. Here is a list of the most significant CTI standards covered in this chapter:

CTI Standard	Description	Endorser
ACT	Applied Computer Telephony	Hewlett Packard
ASAI	Adjunct Switch Application Interface	AT&T
CIT	Computer Integrated Telephony	Digital Equipment Corp.
CSA	Callpath Services Architecture	IBM
SCAI	Switch to Computer Application Interface	Northern Telecom
SCSA	Server API standard byte stream protocol	Dialogic

It is clear from the collaboration of the switch manufacturers with the computer companies on the new CTI standards, that switch makers are experiencing a change in attitude.

There are a number of significant developments, some of which described here, that have some relevance to SCSA. Excepting SCSA, none of the CTI standards handle voice, video, or fax processing, but rather only switching functions. It is not clear whether regular CTI standards will evolve to include these other technologies, or whether host computers will have both CTI and SCSA (Server API) links.

Hewlett Packard ACT

Applied Computerized Telephony

ACT Overview:

Hewlett Packard's entry in the CTI ring is not in the public domain, but ACT supports both Northern Telecom DMS-100, Meridian SL-1 switches. The ACT implementation is leaning towards the ANSI SCAI specification, which is supported and endorsed heavily by Northern Telecom.

No voice processing, fax processing, or other core technology support is currently available, but this may change if the company adopts the Server API specification out of SCSA.

The actual type of link that is most supported with ACT is Northern Telecom's CompuCall (for Central Office Environments) or MeridianLink (for CPE environments). These links use X.25 & BRI ISDN Protocols to connect the switches with the ACT Server.

The typical host used in ACT installations is the HP/3000 or HP/9000. The company supports a Host/Salve or Client/Server model with these classes of computer. Some important HP alliances for ACT are Digital Systems International, Northern Telecom, Brock Control Systems, Coffman Systems, InterVoice, and Siemens/Rolm.

Hewlett Packard's penetration into the CTI market is currently below 5% of all systems installed. The company does have a head start on other host computer manufacturers, however, having spent considerable resources in integrating with both CPE and C.O. environments.

The SCSA architecture and associated SCapi may provide HP with an opportunity to integrate tightly with voice response technology by using the HP/3000 and HP/9000 as the software state controller for SCSA-based VRUs. Since the CIO and NIO (internal) busses are proprietary, HP's quickest access to VRU integration could be via the SCapi byte stream protocol. A PC form-factor VRU could be controlled by the larger UNIX-based HP/9000, which would be similar to what IBM and Tandem are doing.

HP could choose to wait-out the ECMA CSTA and ANSI SCAI outcome before developing additional "one-off" designs with more switch manufacturers.

AT&T ASAI

ASAI Overview:

The ASAI specification is in the public domain and is available from AT&T for a $950 license fee. The link primarily supports the AT&T Definity series of CPE switches, and there is no direct voice processing support.

There are several ways that ASAI hooks-up between the switch and the computer. For high-volume call processing, PRI ISDN lines can terminate on the Definity switch and ASAI messages can be extracted with a Gateway computer and then sent to the host via X.25. This type of arrangement is also supported with BRI ISDN links between the switch and the ASAI gateway computer.

Typical hosts for ASAI applications include the AT&T 3B, IBM, DEC, Stratus, and Tandem machines. Core technology support is minimal, with basic switching the only supported capability at the time of this printing.

AT&T has forged alliances with Tandem, Stratus, EIS (Telemarketing), DEC value added resellers, IBM Business Partners, and Aristacom, for example.

AT&T's ASAI represents perhaps 15% of the approximately 750 overall CTI installations industry-wide. ASAI is limited in its use, although a number of host computers can communicate over the link to a Definity series AT&T switch.

Close to 20% of all AT&T CTI-integrated switches are actually driven by hosts running DEC CIT. Just recently AT&T and Novell announced a joint development initiative for a Novell "Telephony Server NetWare Loadable Module (NLM)" which works via ASAI link to AT&T PC/ISDN card in Novell File Server. It is unclear from this just how significant a role ASAI itself is playing in the development.

In addition, the company just announced the AT&T Passageway Windows API which emulates Merlin Legend phones in PC workstations running Windows. Although the company has not openly supported the SCSA (SCapi) or other CTI standards, pressure from customers may force AT&T to embrace other de facto and de jure standards like the ECMA-backed CSTA and SCSA.

IBM - CallPath Services Environment

CSA Overview:

Callpath Services Environment is not in the public domain, but its supporters are working closely with the ANSI SCAI T1S1.1 working group to iron-out a common set of commands and conventions between the two standards.

In terms of switch support, CSA has an impressive list of "compatible" PBXs. The Call Path /400 and Switch Server/2 options include support for:

- Northern Telecom Meridian 1
- Rolm 9750
- Teleos IRX
- AT&T Definity

In addition, the CallPath CICS package provides support for the AT&T Definity, Rolm 9751, and the Northern Meridian lines.

IBM offers both the DirectTalk/2 & DT/6000 as voice processing support. These systems provide coordinated transfers and voice prompts, outbound dialing, voice recognition, and speech synthesis.

The types of links supported in CSA are as varied as the switch support. Here's a snapshot:

Switch-Side link(s)	**Computer-Side link(s)**	**Adjunct Processor**
ASAI Gateway/DCIU BRI	Token Ring & 802.3	SwitchServer/2.
X.25 & RS-422	Token Ring & 802.3	SwitchServer/2.
X.25 & RS-422	SDLC	SwitchServer/2.
Teleos (Token Ring)	AS/400 (Token Ring)	Direct (no Adjunct)
Meridian X.25/BRI ISDN	X.25	Direct (no Adjunct)
Rolmphone 244PC	802.3, Token Ring, or RS-232	SwitchServer/2

Typical hosts for CSA environments include the AS/400, S/370, PS/2, and RS/6000. Along with voice processing elements, these applications control core technologies such as ASR; Voice I/O, and Speech Synthesis in addition to basic switching functions.

There are a number of alliances that IBM has forged to help push the CSA standard. They include Siemens/Rolm, Teleos, InterVoice, Granada Systems Design, Syntellect, Scott Instruments, Berkeley Speech, AT&T, Northern Telecom, Voicetek, Logica, NPRI, Brock Control, Datacorp, Early Cloud, and Dialogic. IBM is an SCSA Supporter, and has announced that its new products will have closer ties to the architecture.

Even though DEC's CIT is the most popular CTI standard IBM's CallPath Services Environment offers a well-rounded suite of connectivity options. CSA and the associated DirectTalk products provide full NetView and LU6.2 compliance and links to five major switches. With strong support from Siemens/Rolm, Northern Telecom and major VRU vendors, CSA is in the game for the long haul.

DEC CIT

Computer Integrated Telephony

CIT Overview:

DEC's CIT specification is not in the public domain. Licensing is required from DEC, the proceeds of which are used to fund an aggressive programming and research initiative for many switches. This is no small cast, considering the number of switches (individually) supported by CIT:

- Northern Telecom DMS-100
- Northern Telecom Meridian1
- Ericcson MD110
- AT&T Definity Series
- Mitel SX-2000
- Rolm 9750
- Rockwell ACD

There is no VRU support directly from CIT, but a number of integrators have tied-in VRU support from DEC, including DECVoice under VAX Control. Perception Technology PDP-11 based Vocom Series VRUs are also found in CIT environments.

The CIT Applications Interface uses a number of protocols depending on the computer and switching environment. These include DECNet, Ethernet, RS-232, X.25, BRI, and RS-422.

The typical host for the majority of these installations are VAX 4000s running ULTRIX or VMS. DEC has variety of core technology support option from DECVoice and DEC talk including: Voice Store & Forward via DECvoice boards, DECfax Mail for fax, and DECtalk Speech Synthesis.

DEC CIT alliances include Aurora Systems, Sudbury Systems, IOCS, DAVOX, Perception Technology, Northern Telecom, Siemens/Rolm, Mitel, Nabnasset, Target Systems, BellSouth Information Systems, Sandwell, BellCore, Computer Automation, British Telecom, Nynex, and Southwestern Bell.

Seven major switches are supported by DEC CIT, and the list is growing quickly. There are over 300 CIT-based teleservicing installations, by far the most accepted and used standard today.

DEC provides a compelling argument for how a non-telecom company can become a major player in teleservicing. Despite the fact that the company does not manufacture switches, its "call center interop" strategy is pulling through many VAXs and MicroVAXs.

Along with IBM, UNISYS, and Tandem, DEC has made serious investments in the area of voice processing. The DECTalk line will undergo radical architecture changes as a result of the new Alpha Chip architecture, and DEC will probably continue to leverage its hooks into the call center with CIT software. The company has pledged to migrate large service bureaus and telcos onto the new RISC-based platform. CIT is the "glue" that will help DEC to do this.

SCAI

Meridian Link / CompuCall

SCAI Overview:

The Switch Computer Application Interface (SCAI) is the ANSI (American National Standards Institute) version of ECMAs CSTA CTI standard. SCAI is being standardized by the ANSI T1S1.1 Working Group. The SCAI interface is more or less in the public domain as a result of Northern's activity in the ANSI Working Group. The link supports both the DMS-100 & Meridan 1 line. Currently, there is no VRU Support per se, but Northern does sell a variety of separate voice processing systems including the Meridian Mail and StartTalk voice messaging products.

The SCAI link uses either X.25 or Basic Rate ISDN protocols for hosts such as the Tandem CLX, HP32044A (ACT Server), HP 3000/9000, and IBM AS/400. LAPB is used for the VAX line of computers from DEC.

Currently, the link supports only switching functions, but Northern has other core technologies handled in its Norstar Hybrid via the Access product and associated PC integration cards.

Northern counts the following companies as ASAI supporters and users: Ameritech Information Services, Appintec (TeleMAGIC), Bell South Information Systems, Coffman Systems, DSI (VoiceLink), IBM CSA Partners, Datacorp Business Systems, and NPRI.

Northern Telcom has been the most active switch manufacturer in the standards initiative and offered much of the specification to ANSI as the company began to implement trials of its own versions of the proposed standard: Meridian Link and CompuCall.

CompuCall is serves as a real-time link between host computers and the DMS-100 central office. This is the first instance of a central office switch manufacturer seriously getting involved in CTI. In fact, it was late in 1990 when IBM and Northern Telecom jointly trailed the capability with an AS/400 on a college university campus.

The interface was completed through adjunct links between IBM's CSA (Call Path Services Environment) and CompuCall. Much in the same fashion the HP ACT Server acts as the "protocol converter" between CompuCall and the HP/3000/9000. Although not as successful as DEC CIT, Northern's flavor of CTI is likely to go a long way considering the open nature of the specification and the company's standards activities.

CTI and SCSA -
"The ultimate mate between host and switch and VRU"

The SCSA specifications are available from Dialogic for a $125 documentation fee. The "switch" support is for any device that uses SCbus, SCxbus, or PEB (PCM Expansion Bus). This is because the Server API specification is a means to drive the switching (and core technology capabilities) of any signal computing node that conforms to the other end of the interface specification.

Currently, the Server API specification lives in a product called **AppServer**, which is a Dialogic software product. This product is an instantiation of the Server API specification in its first phase. VRU support such as on hook, off hook, record, playback, and DTMF commands are handled today. In the second and third releases, advanced switching, timeslot bundling, fax, and voice recognition will be handled. In addition, various computing models will be supported in the third release including self-hosted, Host/Slave, and Client/Server.

The SCSA-based Server API link uses a byte stream protocol over asynchronous RS232 or Ethernet - TCP/IP links. the typical host for the Server API are various Intel 386/486 PCs, Tandem non-stops, and Sun Servers.

Although not in full swing, and still officially in the "Concept" stage, the Server API Functional Description was the basis for Tandem and Vicorp International's current VRU offerings. This includes the Tandem 2400 VRU and the BetexESP package, productized by Tandem and ViCorp International, respectively.

The Signal Computing System Architecture was announced by Dialogic Corporation in early March of 1993. SCSA has garnered over 100 cross-industry endorsements in the areas of computing, telecom, and call processing. The support of the architecture comes form companies like IBM, Siemens, Analog Devices, VLSI, NEC, Northern Telecom, Vicorp International, Tandem, and Harris DTS.

The proposed standard seeks to ratify the common elements found in all communication systems and define specifications for interoperability between them at virtually every level of the OSI 7-layer model.

At the physical and data link layer, a high-speed PCM/TDM bus has been defined which carries as many as 2,048 timeslots of voice, video, or regular data. The bus uses HDLC signaling for out-of-band control between core technologies sharing the bus.

At the highest level, a host independent byte stream protocol, which can be implemented asynchronously over RS 232, or via 802.3 has been defined which allows host computers to issue real-time commands to SCSA-based VRUs. In Phase II of Dialogic's "AppServer" (the company's implementation of SCapi), both switching and collateral technology support of fax and voice recognition will be available. SCSA could re-define interoperability in teleservicing.

Crossing distribution and development channels

The table below details the traditional distribution for messaging, IVR, PC's and other communications gear. There are significant opportunities for "cross-over" between the technologies themselves as a result of the CTI standards, and also opportunities for distribution. Interconnect companies will sell more computers. Systems Integrators will sell more telephone systems. And sooner or later (probably sooner), telecom gear will become commoditized to the point where it can be purchased in local stores and catalogues, just the way we do now with PC hardware and software.

Manufactured Item	Typical Channel	Type User
Voice Messaging	Voice Processing Manufacturer, Switch Manufacturer, Interconnect, Telco	Commercial, Residence
PCs	Computer Wholesalers, LAN installers, Direct Sales, Systems Integrator	Commercial
Switches	Switch Manufacturer, Interconnect, Telco	Commercial
Interactive Voice Response	Manufacturer Direct, Systems Integrator	Commercial
Telemarketing Gear	Manufacturer Direct, Systems Integrator	Call Center Businesses

CHAPTER 10 - The SCSA Family of Developers

At this printing, the number of individuals and companies who are holders of the SCSA technical documentation is well over 100. This list is growing quickly, just as the industry of call processing is growing.

Perhaps one of the biggest challenges in melding the many technologies that are addressed by SCSA is the complexity of synthesizing input from users, developers, and manufacturers. This is difficult because as much as technology leaders want to be perceived as benevolent, they also want to cut through all of the standards-setting politics and get some good development under their belts so they can sell products and services.

Dialogic wants to be perceived as both benevolent and as a technology pace-setter. To that end, the company has developed a straight-forward means to solicit input, suggestions, and feedback from SCSA constituents. This method is manifest in both a users group, and a collection of associated "working groups." There are several forums for both the user group and working groups to communicate, and this chapter discusses some of those proposed methods.

Dialogic has taken a "Microsoft" approach to soliciting input for standards under SCSA. That is, that by working closely with recognized experts in certain areas of technology, Dialogic develops a draft specification and then lets it out for comment to a group of concerned constituents and experts. These comments are solicited over a short period of time, synthesized, and then the company implements the proposed change with the guidance of members of the affected working group.

This is a departure from the development of draft international standards, which are developed in committee and distributed over a broad audience and several years. Although Dialogic is not averse to contributing some portions of the SCSA specification into international standards bodies, it is clear that the company wishes to take the de facto route in order to expedite solutions into the marketplace.

SCSA User Group

Dialogic conducts the efforts of the SCSA constituency through a newly-formed SCSA User Group. The way the group interacts with Dialogic is pictured in **Figure (CONDUCT0.DRW).** The purpose of the group is to discuss of issues and make proposals for the standardization of interfaces for multiline call processing systems. As the role of call processing expands and plays a larger role in day-to-day business, the purpose of the group will probably change.

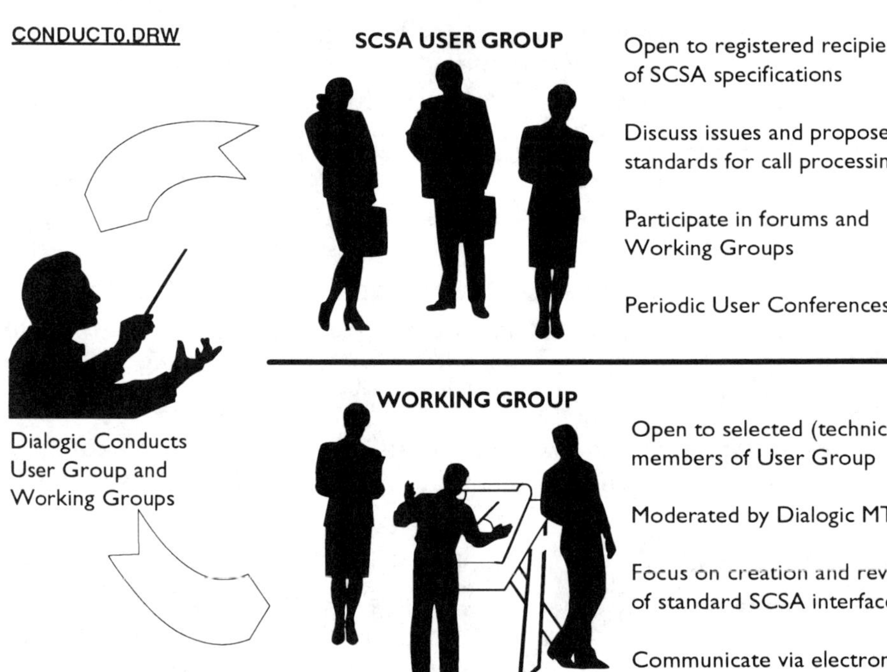

User Group membership is open to all registered recipients of the official SCSA specifications which are available through Dialogic. The company says that it's all right to copy the document and distribute it, however, only registered users get updates and notification of user group meetings, conferences, and change proposals.

As an SCSA User group member, individuals are encouraged to participate in forums. Forums are discussion groups that are moderated by User Group members. These forums are open to both SCSA User Group members and other selected participants in order to focus on key technology issues. The idea is to bring together interested parties from different companies and disciplines in order to share ideas across a diverse knowledge base. Forums under consideration are listed below.

- Voice
- Fax
- ASR
- TTS
- Video
- Network Interfaces
- Datacomm
- SCbus and SCxbus
- Operating Systems
- Testing and Compliance

Periodic user group conferences are proposed as well. These face-to-face meetings will provide instant feedback to Dialogic on timely issues, and also give user group members a chance to meet one another and discuss business in general including possible co-marketing and joint development opportunities.

SCSA Working Groups

Working groups are a sub-set of the SCSA User Group. A working group is open only to selected members of the User Group and is moderated by Dialogic MTS (members of the technical staff). Working groups focus on creation and revision of standard SCSA interfaces and are typically very small in size. the idea is to quickly develop and ratify proposed interfaces and revisions in order to speed solutions to the user group at large. Because of this "rapid development" charter, the members of the working group are chosen on a very selective basis by Dialogic MTS. Only absolute technical experts in a chosen field are used to form a group, since the non-technical issues and business issues are typically handled by the User Group.

Working group members communicate via electronic messaging over Tymnet and Internet.

Working Group Specification Process

These small groups first define a technology-based concept and the draft requirements based on User Group input. The working group provides input and feedback on specification revisions and then circulate "final" specification to the Users Group for feedback. Once the feedback loop is closed, the working group incorporates new specifications into the SCSA technical manual.

The communication vehicles for the SCSA User Group are pictured in **figure (VEHICLE.DRW)**. The lines of communication that have been set up include regular voice communication in the form of point-to-point phone calls and teleconferences, fax communication, user group meetings, and e-mail. The e-mail method is both regular e-mail via Internet and also e-mail serviced by an SCSA conferencing server called "***scsa***."

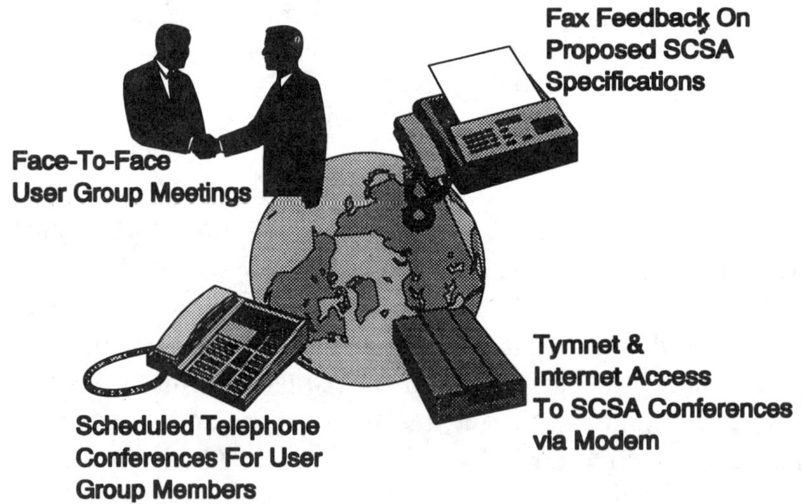

Conferencing Server Connect Process

Tymnet is a global X.25 network which users can access in any country. Most accesses to Tymnet are achieved using a local telephone number, and

Tymnet carries the rest of the communication through packet switches to provide the rest of the connection. To log onto ***scsa*** using Tymnet, a user must have a Tymnet user name and password.

Once logged into Tymnet, the user will be presented with ***scsa's*** login prompt. At this point, the member's ***scsa*** user name and password are entered. The user is then connected to ***scsa*** and the first menu appears on the screen.

Dialogic will be billed for all Tymnet access to ***scsa***, so the company bills SCSA user group members who use Tymnet periodically for their access time. This is not the case if a user dials direct to a set of ***scsa*** modems, because there are fewer costs incurred by Dialogic.

In addition to the ***scsa*** conferencing server, a regular E-mail feedback channel to Dialogic MTS can be accessed through Internet as follows:

> ***<name code>@dialogic.com***
>
> where ***<name code>*** is the person's first initial followed by a period and then their last name

The ***scsa*** conferencing server access is via Tymnet and Internet. Logon access may change, but is initially assigned the following access code:

> ***usergroup@scsa.dialogic.com***

Conferencing Server

The ***scsa*** conferencing server is a microcomputer. It uses commercial e-mail and conferencing software. It can be configured for a variety of conference options and levels of security for user access via user account profiles. Standard processes for sign-up, sign-on, and conference notification are being finalized.

The server is a 486SX running Interactive UNIX v3.0.1. The unit is currently installed at Dialogic's Parsippany, NJ headquarters and is connected to the Dialogic (private) global e-mail network and to Tymnet with two 14.4 Kbps modems for dial-up access. The Internet access is still being finalized, however, anyone at Dialogic can be contact via Internet as stated above.

The unit is set-up for conferences by pre-assigning a moderator to each conference. The system will support conferences and threads within a conference. It allows users to upload to or download from any thread in the conference that you are in. Threads are established by the "subject" line in the message. Any user can start a new thread by typing in a new subject.

The ***scsa*** conferencing server also provides a means of archiving files which can be downloaded by a user. Restrictions may be put on users as to what files are visible to them, depending on their user ID. Restrictions may also be put on users as to what conferences they can read and to which they can write, based on their user ID.

Moderators

Conference moderators are responsible for replying to conference messages when necessary and for monitoring the messages which are sent to the conference to make sure they are productive. Conference moderators have read/write access to all files in a conference. They will be able to edit and/or delete files as deemed necessary for their conference.

File Archival

The ***scsa*** computer will provide access to SCSA specifications. Any specification that has reached the evaluation level will be accessible by all user group members for feedback. Specifications still at the working group level are accessible only by members of that working group.

Specs will be stored on ***scsa*** in PostScript format. The idea behind archiving the specifications in this fashion is to allow user group members to download specifications on demand to their own systems. A mechanism for uploading documents will also be available.

Conferences

There will be a marketing and technical conferences will be sponsored for SCSA User Group members. The working group conferences will be accessible only by active members of the working group. The table below provides a list of proposed conferences, moderating groups, and member types.

Conference	Moderating Group	Type
SCSA Marketing	Marketing	user group members
SCSA Technical	Development Staff	user group members
SCbus & SCxbus	Engineering; Systems	working group members only
SCfirmware	Engineering; Firmware	working group members only
SCdpi	Engineering; Software	working group members only
SCapi	Engineering; Software	working group members only
VR DPI	Engineering; Software	working group members only
TTS DPI	Engineering; Software	working group members only
Fax DPI	Engineering; Software	working group members only

Who will be involved in the SCSA User Group and associated forums?

Although it's hard to predict exactly who will be most involved and to what extent, the specification holders and supporters of the SCSA architecture are a good bellwether for types of companies who will be involved.

The company contact data below is a partial list of SCSA supporters and specification holders. The list includes technology experts, service bureaus, large international companies, software tool providers, computer companies, and telephone switch companies. In short, there is significant cross-industry representation in this group of supporters such that a comprehensive view of user needs is anticipated.

A+ Systems Corporation
14 Marshall Lane
Weston, CT 06883-1230
Phone: 203-221-8181
Fax: 203-221-8179
Contact: Ami Dabush

ACCI
501 Office Center Drive
Suite121
Fort Washington, PA 19034
Phone: 215-540-9377
Fax: 215-540-9383
Contact: Tony Agostinelli

Acsys S.A.
6 Rue des Coutures
Z.I. Sud F-77200
Torcy, France
Phone: 33-1-6017-6690
Fax: 33-1-6017-3755
Contact: Jean-Noel Charpiat

Active Voice Corporation
2901 Third Avenue
Seattle, WA 98121
Phone: 206-441-4700
Fax: 206-441-4784
Contact: Robert Greco

Aculab, Ltd.
Unit A Station Approach
Leighton Buzzard, Berks
LU7 7LU United Kingdom
Phone: 44-525-371393
Fax: 44-525-381284
Contact: Alan Pound

ADX Microsys Corporation
5959 W. Century Blvd.
Los Angeles, CA 90045
Phone: 310-642-0250
Fax: 310-642-0252
Contact: Barbara Hirsch

Aerotel Ltd.
5 Hazoref Street
Holon 58856, Israel
Phone: 972-3-559-3222
Fax: 972-3-559-6111
Contact: Rafi Katz

AMTELCO
4800 Curtin Drive
McFarland, WI 53558
Phone: 608-838-4194
Fax: 608-838-8367
Contact: Carl Pigott

Analog Devices
3 Technolgy Way
Norwood, MA 02062
Phone: 617-461-4010
Fax: 617-461-4400
Contact: Tim Counihan

APEX Voice Communications
14900 Ventura Blvd.
Suite 200
Sherman Oaks, CA 91403
Phone: 818-379-8400
Fax: 818-379-8410
Contact: Ben Levy

Applied Voice Technology
11410 NE 122nd Way; PO 97025
Kirkland, WA 89083
Phone: 206-820-6000
Fax: 206-820-4040
Contact: Thomas W. Beehan

Atlas Telecom
4640 SW Macadam Ave., Suite 96
Portland, OR 97201
Phone: 503-228-1400
Fax: 503-228-0368
Contact: Bill Draeger

AudioFAX, Inc.
2000 Powers Ferry Road; Suite 200
Marietta, GA 30067
Phone: 404-933-7600
Fax: 404-933-7606
Contact: Charles M. Brewer

Berkeley Speech Technologies
2246 Sixth Street
Berkeley, CA 94710
Phone: 510-841-5083
Fax: 510-841-5093
Contact: Sara S. O'Malley

Biscom, Inc.
85 Rangeway Road
Billerica, MA 01821
Phone: 508-670-5521
Fax: 508-671-3970
Contact: Steve Pytka

BMC Group, Inc.
785 English Court
Lawrenceville, GA 30244
Phone:404-985-6609
Fax:404-978-9739
Contact: Bill Valles

Boston Technology
100 Quannapowitt Parkway
Wakefield, MA 01880
Phone: 617-246-9000
Fax: 617-245-5322
Contact: Paul DeLacey

Brite Voice Systems
7309 East 21st Street North
Wichita, KS 67206-1083
Phone: 316-652-6543
Fax: 316-652-6800
Contact: Don Mounday

C-T Link
165 Orchard Avenue
Weston, MA 02193
Phone: 617-737-1277
Fax: 617-737-7796
Contact: James Burton

CadCom Telesystems, Inc.
1296 Atlanta Road
Marietta, GA 30060
Phone: 404-422-3600
Fax: 404-590-5113
Contact: James Tabb

Call Completion Technologies
6230 Fairview Road; Suite 200
Charlotte, NC 28210
Phone: 704-527-8888
Fax: 704-365-0498
Contact: Donald Beck

Cascade Technologies, Inc.
1430 Broadway
New York, New York 10018
Phone: 212-768-7380
Fax: 212-768-7806
Contact: Neil Swanson

CNET
2 Rue de Tregastel
PO F-22301
Lannion Cidex, France
Phone: 33-96-05-3088
Fax: 33-96-05-1412
Contact: Andre Soubigou

Computer & Communications Co. Ltd
St. John's Innovation Center
Cowley Road
Cambridge, United Kingdom
Phone: 44-223-423-562
Fax: 44-223-420-709
Contact: Maurice Jones

Computer Talk Technology
225 East Beaver Creek Road
Richmond Hill, Ontario,Canada
Phone: 416-882-5000
Fax: 416-882-5501
Contact: Mandle Cheung

Cypress Research
240 East Caribbean Drive
Sunnyvale, CA 94089
Phone: 408-752-2700
Fax: 408-752-2735
Contact: Paul Hurley

Datakinetics, Ltd.
15 A Salisbury Street
Fordingbridge, Hampshire
England SP6 1AB
Phone: 425-655-050
Fax: 425-655-075
Contact: David Fletcher

Dialogic Corporation
300 Littleton Road
Parsippany, NJ 07054
Phone: 334-1268
Fax: 334-1257
Contact: Terry Henry

Digital Equipment Corporation
Digital Drive MKO1-2/HO1
Merrimack, NH 03054-9501
Phone: 603-884-4621
Fax: 603-884-2607
Contact: Anil Kapoor

Efrat Future Technology, Ltd.
Yigal Alon 110
Tel-Aviv, Israel 67891
Phone: 972-3-513-1331
Fax: 972-3-513-1309
Contact: Sachi Gerlitz

Elan Informatique
4 Rue Jean Rodier
31400 Toulouse, Fance
Phone: 33-61-36-0777
Fax: 33-61-36-0770
Contact: Bertrand Chauvet

Enhanced Systems, Inc.

6961 Peachtree Industrial Blvd.
Norcross, GA 30092
Phone: 404-662-1503
Fax: 404-242-1630
Contact: Fereydoun Taslimi

Equal Access Corporation
11150 Santa Monica Blvd.Suite 360
Los Angeles, CA 90025
Phone: 310-445-9100
Fax: 310-445-5180
Contact: Farzad Mobin

Executone Information Systems
6 Thorndal Circle
Darien, CT 06820
Phone: 203-655-6500
Fax: 203-656-5778
Contact: Michael Yacenda

Expert Systems, Inc.
1301 Hightower Trail
Suite 201
Atlanta, GA 30350
Phone: 404-662-7575
Fax: 404-587-5547
Contact: Carlton Carden

Global Communications, Ltd.
211 Picadilly
London W1V 91D; U.K.
Phone: 44-71-4120214
Fax: 44-71-4131032
Contact:Michael Margolis

Harris Digital Telephone Systems
PO Box 1188
Novato, CA 94948-1188
Phone: 415-382-5083
Fax: 415-883-1626
Contact: Ajoy Khandheria

I+D Telefonica
Emilio Vargas 6
28043 Madrid, Spain
Phone: 34-1-337-4008
Fax: 34-1-337-4322
Contact: Alejandro Acero

IBM
Mail Point 183
Hursley Park, Winchester
Hampshire SO21 2JN; U.K.
Phone: 0962-884-433
Fax:0962-849-268
Contact: John Mears

Infologue, Inc.
270 Lancaster Ave.
Suite B1
Frazer, PA 19355
Phone: 215-647-7584
Fax: 215-651-0914
Contact: Kasim Mun

Infovox AB
PO 2069
Phone: 46-8-764-35-00
Fax: 46-8-735-78-76
Contact: Staffan Prior

Intellivoice Communications, Ltd.
32605 West Twelve mile Road
Suite 300
Farmington Hills, MI 48334
Phone: 313-488-0180
Fax: 313-488-2101
Contact: Jim Zyperski

Interact, Inc.
770 North Corner Blvd.
Suite 401
Lincoln, NE 68505
Phone: 402-464-8786
Fax: 402-464-8850
Contact: Lynn McKee

International Telesystems, Ltd.
18A Broadwalk, Pinner Road
North Harrow, Middlesex
HA2 6ED United Kingdom
Phone: 44-81-424-0527
Fax: 44-81-424-0528
Contact: Steven Ekbery

Inter-Tel Equipment, Inc.
6505 West Chandler Blvd.
Chandler, AZ 85226
Phone: 602-961-9000
Fax: 602-961-1370
Contact: Thomas Parise

Kreutler
Sophienstrasse 96-104
W-7500 Karlsruhe 1; Germany
Phone: 0721-8508-0
Fax: 0721-8508-90
Contact: Eberhard Schuler

Lernout & Hauspie
800 West Cummings Park
Suite 3100
Woburn, MA 01801
Phone: 617-932-4118
Fax: 617-932-9209
Contact: Bart Verhaeghe

Logica Cambridge Ltd.
Betjeman House
104 Hills Road, Cambridge
CB2 1LQ United Kingdom
Phone: 44-223-66343
Fax: 44-223-322315
Contact: Jeremy Peckham

Melita International
2935 Bankers Industrial Dr. NW
Doraville, GA 30360
Phone: 404-368-8076
Fax: 404-446-2404
Contact: Mani Subramanian

Microlog Corporation
20270 Goldenrod Lane
Germantown, MD 20874
Phone: 301-428-3227
Fax: 301-916-2474
Contact: Dennis Grundy

NEC America
1525 Walnut Hill Lane
Irving, Tx 75038
Phone: 214-518-3802
Fax: 214-518-3990
Contact: Yoshihiko Katsura

Northern Telecom
565 Marriott Drive
Suite 300
Nashville, TN 37210
Phone: 615-391-2250
Fax: 615-391-2202
Contact: Curtis Kinsman

N-SOFT
52 Ave. due 8 Mai 45
Sarcelles, 95200; France
Phone: 33-1-39-92-23-44
Fax: 33-1-39-92-44-46
Contact: Eric Messika

Parity Software Development
25 Stillman Street, Suite 106
San Fransisco, CA 94107
Phone: 415-931-8221
Fax: 415-546-7329
Contact: Bob Edgar

PCBX Systems
3730 South Susan Street, Suite 100
Santa Ana, CA 92704
Phone: 714-668-1180
Fax: 714-668-0215
Contact: John Alkire

Probe Research, Inc.
8077 Hihn Road
Ben Lomond, CA 95128
Phone: 408-336-5123
Fax: 408-336-2171
Contact: Harry Schwedock

The Renaissance Group
950 South Bascom Ave.
Suite 1111
San Jose, CA 95128
Phone: 408-287-9590
Fax: 408-287-9590
Contact: John Appler

Rochelle Communications
4030 - 1 West Braker Lane
Austin, TX 78759
Phone: 512-794-0088
Fax: 512-794-0908
Contact: Gilbert Amine

Sandwell, Inc.
2575 Cumberland Parkway
Atlanta, GA 30339
Phone: 404-432-9450
Fax: 404-432-9092
Contact: Greg Mohl

Scott Instruments
1111 Willow Springs
Denton, TX 76205
Phone: 817-387-9514
Fax: 817-566-3174
Contact: Ric Adams

SDX Business Systems, Ltd.
Sterling Court 15-21 Mundells
Welwyn Garden City
Hertfordshire, AL7 1LZ; U.K.
Phone: 44-707-392200
Fax: 44-707-376933
Contact: Kuir Wink

Siemens Corporation
2191 Laurelwood Road
Santa Clara, CA 95054
Phone: 408-980-7912
Fax: 408-980-4510
Contact: Paul Morrison

SpeechSoft, Inc.
32 Manners Road
Ringoes, NJ 08551-1714
Phone: 609-466-1100
Fax: 609-466-0757
Contact: Morris Neumann

SRX
3480 Lotus Drive
Plano, TX 75075
Phone: 214-985-2614
Fax: 214-985-2606
Contact: Jeff Box

Summit Telecom
11150 Santa Monica Blvd., Suite 360
Los Angeles, CA 90025
Phone: 310-445-9100
Fax: 310-445-5180
Contact: Farzad Mobin

Syntellect, Inc.
15810 North 28th Ave.
Phoenix, AZ 85023
Phone: 602-789-2800
Fax: 602-863-6561
Contact: Lane Dixon

T1 Systems
1658 Cole Blvd.
Boulder, CO 80401
Phone: 303-237-0410
Fax:303-237-0526

Tandem Computers, Inc.
19191 Vallco Parkway
Cupertino, CA 95014-2525
Phone: 508-369-8146
Fax: 408-285-1800
Contact: Chuck Buffum

Telecorp Systems, Inc.
100 Holcomb Woods Parkway.
Roswell, GA 30076
Phone: 404-587-0700
Fax: 404-587-0589
Contact: Steve Nussrallah

Teledirect International
5510 Utica Ridge Road
Davenport, IA 52807
Phone: 319-355-6440
Fax: 319-355-4890
Contact: Glenn Thorne

Telephone Response Technologies, Inc.
1624 Santa Clara Drive
Suite 200
Roseville, CA 95661
Phone: 916-784-7777
Fax: 916-784-7781
Contact: Edwin Margulies

Tern Systems
168 Stone Root Lane
Concord, MA 01742
Phone: 508-369-8146
Fax: 508-369-8146
Contact: Walt Tetschner

U.S. Telecom International
211 Main Street
Suite 401
Joplin. MO 64801
Phone:417-781-7000
Fax: 417-623-2963
Contact: David King

Vicorp International Services
28-30 Bd de la Cambre
b-1050 Brussels
Belgium
Phone: 32-2-646-5190
Fax: 32-2-646-5964
Contact: Daniel G. Schultz

Vicorp
399 Boyleston Street
Boston, MA 02116
Phone: 617-536-1200
Fax: 617-536-6647
Contact: Stuart Patterson

The Vmail Company, Ltd.
Sherfield Building
Imperial College
London, England SW7 2AZ
Phone:44-71-589-1036
Fax: 44-71-589-9380
Contact: Henry Hyde-Thompson

Voice Control Systems
14140 Midway Road, Suite 100
Dallas, TX 75244
Phone: 214-386-5555
Fax: 214-386-0300
Contact: Peter Foster

Voice Data Systems B.V.
Newtonlaan 211
3584 BH Utrecht
Netherlands
Phone: 31-30 518888
Fax: 31-30-517-517010
Contact: Henk van den Broeck

Voice Information Systems, Inc.
24 North Marion Ave.
Suite 353
Bryn Mawr, PA 19010
Phone: 215-747-5035
Fax: 1-800-234-3948
Contact: Andrew Michalik

VoicePlex Corporation
1425 Greenway Drive, Suite 250
Irving, TX 75038
Phone: 214-550-0900
Fax: 214-550-0458
Contact:Sohail Sattar

Voice Processing Corporation
One Main Street
Cambridge, MA 02142
Phone: 617-494-0100
Fax: 617-494-4970
Contact: Desmond Pieri

Voice Technologies Group
2350 North Forest Road
Buffalo, NY 14068
Phone: 716-689-6700
Fax: 716-689-6800
Contact: David M. Straitiff

VoiceSmart Corporation
2116 North Central Road
Fort Lee, NJ 07024
Phone:201-592-0505
Fax: 201-592-6962
Contact: Carol L. Stone

Voicetek Corporation
19 Alpha Road
Chelmsford, MA 01824
Phone:508-250-7921
Fax:508-250-9378
Contact: Jan Bergeron

CHAPTER 11 - SCSA Developers and What They Do

No one vendor can claim total compliance with SCSA, because it is an architecture which allows companies with differing competencies to comply in their respective area of expertise. For this reason, compliance with SCSA is done either at the bus level, the API level, or the firmware level, for example. The companies highlighted in this chapter have decided to support the SCSA initiative along with over one hundred other companies. Of the designs illustrated, some are either already implemented or are planned.

What's encouraging is just how "open architected" many companies already are. SCSA provides an infrastructure of common specification and working forums for these developers. The following companies will be profiled here because they are widely recognized, and because both telephone switch and computers are represented by these organizations:

- Tandem Computers
- IBM
- Harris (Digital Telephone Systems Division)
- NEC
- Northern Telecom

Tandem Computers

"The SCSA architecture is ideal for client/server voice processing applications. Standardized interfaces for offboard processing will drive a wider variety of solutions for our industry." -Chuck Buffum, Tandem Computers

Tandem enjoys some 70% of the ATM (Automatic Teller Machine) market today. According to Mr. Buffum, the company's fault tolerant line of computers is also used in 9 of the top 10 EFT (Electronic Funds Transfer) networks.

Tandem computers are used in the OSSs (Operational Support Systems) of all of the world's 32 largest telephone companies. Since Tandem is a leader in the world of on-line transaction processing, call processing and call center-based transaction processing is a natural fit for the company.

Tandem ties the call center together by applying their transaction processing expertise to telephone automation. This is achieved through a computerized Automation Controller, which commands voice response functions over an RS-232 control link. The Automation Controller also handles telephone switch commands via the Control Link Manager. In addition, a Tandem host can control outbound dialing and other voice processing functions.

This is enabled, in part, because of Tandem's pioneering work with both ViCorp International and Dialogic. An early implementation of the SCSA-based ServerAPI specification is what is used in this scenario to control the VRUs. The actual product is called "AppServer" from Dialogic and the Host control software running under Tandem Guardian is "BETEX-ESP" from ViCorp.

These are the basic building blocks for the Tandem Call Applications Manager (CAM), which the company regards as the very heart of any modern call center. CAM acts as a "traffic cop" by controlling inbound & outbound telephone traffic in the call center. The CAM package offers simplified application development and allows for the integration of VRU functions with switch-to-computer links that support multiple PBXs and ACDs. Figure **(TANDEM01.DRW)** illustrates the way all of the pieces work together in a typical call center application with Tandem's solution.

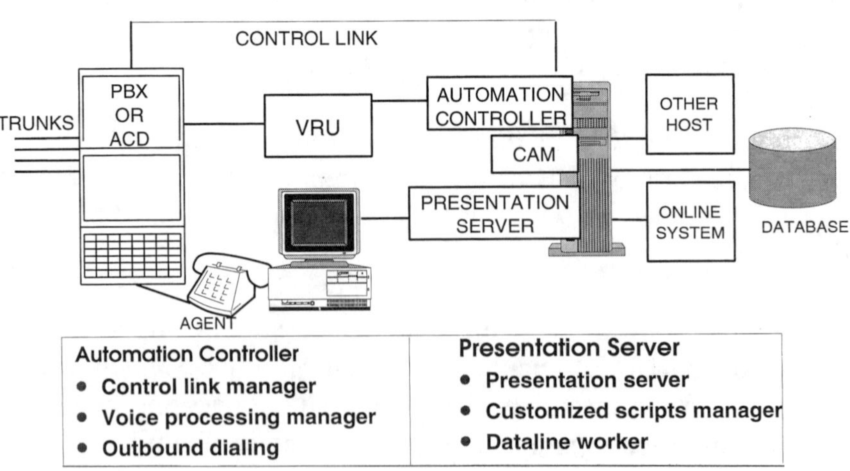

Tandem's Voice Response Unit comes in two models: the 2400 VRU and 4800 VRU. They are both Dialogic-based in the sense that the Tandem VRUs house Dialogic cards and the UNIX-based AppServer technology. The VRUs are interactive voice response units that are host-controlled from the Tandem computer. The models handle 24 or 28 lines of simultaneous IVR, respectively, and by networking units together, a system is scalable to hundreds of lines.

Both 2400 and 4800 VRU models can be remotely connected to a host processor via asynchronous communications lines, and offer standard analog connections to ACDs and PBXs.

The 2400 VRU/4800 VRU enables coordinated transfer of voice and data via Call Applications Manager (CAM), and the support of geographically distributed call centers. Since the host is fault tolerant, this package will be popular for mission-critical service environments.

Tandem Product Brief:

Processor	(used in)	Op. System	Resiliency	Form Factor
Intel 386/486	2400/4800 VRU	UNIX/SCO	Commercial	ISA/PC
Proprietary	XTP; CLX; VLX; Cyclone	Guardian/UNIX	Fault Tolerant	VME

Tandem Typical Application Environments

Application	*Messaging*	*Audiotex*	*Fax S&F*	*Host IVR*	*Call Center*
CPE				☒	☒
NTWK	☒	☒	☒		

Tandem HOST & CTI Interfaces:

CAM (Call Applications Manager), X.25.
SCapi (SCSA) Dialogic AppServer/UNIX (RS-232).
bX.25 for Rockwell Transaction Link.
X.25 / Meridian Link. AT&T DCIU

Tandem Interop Factor:

High - can control any Dialogic-based VRU running AppServer UNIX.

Northern Telecom Merdian 1, and AT&T Definity switches.
6100 I/O & 5600 card provide various I/O.

Tandem Alliances:
U.S. West, Cable & Wireless, AT&T, TTSI,
SCSA supporter, National Data Corp.

Tandem Used In What Systems:
2400 VRU, 4800 VRU, high-end OLTP, telco enhanced
services environments, intercept treatment, ACD host control.

Call Processing Background:

Tandem has had plenty of experience in the voice and call processing area. The company once attempted its own manufacture of voice processing cards back in 1986, but the project was put on the back burner in favor of co-marketing relationships with companies like Votrax, IOCS (VoiceNet/ISFEP), and Voicetek (Wye Tech Software).

Recent developments point to some forward thinking on Tandem's part: It's support of ViCorp International's BETEX-ESP Service Creation Environment, and it's OEM deal with Dialogic for voice cards in Tandem's PC-based 2400 and 4800 series VRU, which are being marketed by Tandem's Call Center Marketing Group.

IBM

> ***"SCSA (provides) opportunities for the DirectTalk/2 Open Architecture including support for industry technologies and expandability via user actions, clients, and servers." - John Mears, IBM Computer Aided Telephony Systems***

The Computer Aided Telephony Systems group is part of IBM's networking systems line of business. This group has a worldwide focus and is headquartered in Hursley Park in the U.K. The company has a full compliment of call processing and voice processing products, and is absolutely expert when it comes to network integration.

The CallPath products, and especially the DirectTalk/2 (DT/2) line are manifestations of a whole new vision on the part of IBM. The DT/2 line is part of the CallPath Services Architecture, which handles both traditional SNA/3270 implementations as well as more modern Client/Server architectures.

Computer Aided Telephony Systems provide worldwide coverage. This is due in part to the international presence of IBM itself, and also due to Dialogic's very aggressive international approvals campaign. Together, the companies have PTT-approved products that are "homologated" (approved) worldwide.

The company does have proprietary implementations of call processing technology including the DT/6000, however, the focus and forward approach of the company is to support open standards, thus their support of SCSA.

John Mears says the his company wants to "enable rapid evolution through open architecture." He plans to do this by supporting heterogeneous switches and host computers, and to do it in multiple languages. According to Mears, the CallPath DirectTalk/2 product line fills the bill, because it supports more than 20 spoken languages, and is "homologated" in more than 20 countries. The line sports an extensible architecture, which takes advantage of a client/server paradigm.

To a large degree, IBM acts in the role of a consulting house and systems integrator. Yes, there's pull-through of their "big and little iron," but on the most part, the company is focusing on integration solutions. Figure **(IBM.DRW)** illustrates the DT/2 client/server architecture and the way the company hooks-up to host computers and switches. This is a fairly high-level diagram, and the specific implementation details of an installation vary widely.

For example, a DT/2 customer may have multiple voice development clients that sit on the same LAN as a SwitchServer/2 computer. The Switch Server is a CTI device which acts as a protocol converter between the service creation and run-time environment and the actual PBX or ACD itself. The same LAN-based call processing system can also be attached to a variety of IBM hosts, including 3090s and AS/400s, for example.

IBM.DRW

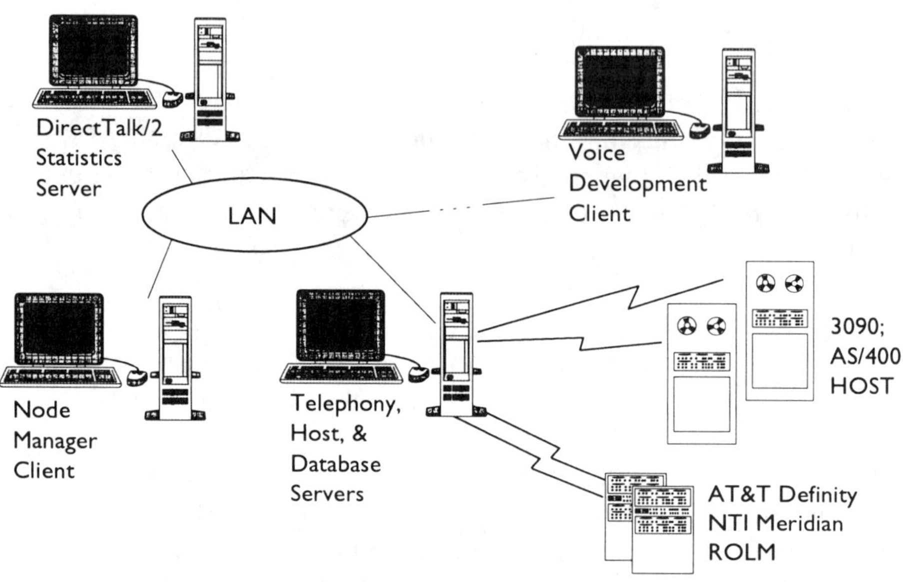

There's a number of options that allow NetView and LU6.2 peer-to-peer support between the Dialogic-based DT/2 units and the client's computer network. Add this to full support of AT&T, Northern Telecom, and Rolm switches, and you've got a real international powerhouse on your hands...

It's likely that IBM will fist support SCbus-based hardware products that will be housed in its PCs, and then later support API level specifications under SCSA. The company has a keen interest in voice recognition and international language support, so there's plenty of room for OEM relationships and development deals.

IBM Product Brief:

Processor	(used in)	Op. System	Resiliency	Form Factor
RISC/6000	DirectTalk/6000	AIX; AIXWindows	High	MCA/PC
Intel 386	PS/2 - DT/2	OS/2	Commercial	ISA/PC
IBM	AS/400; S/3x	AS/IMS	High	Propietary
IBM	S/370	CICS/MVS	High	Propietary
IBM	S/390	CICS/MVS	High	Propietary

IBM Typical Application Environments

Application	*Messaging*	*Audiotex*	*Fax S&F*	*Host IVR*	*Call Center*
CPE		☒		☒	☒
NTWK	☒	☒		☒	☒

IBM HOST & CTI Interfaces:

CallPath/2 & SwitchServer provide AT&T ASAI gateway to PS/2s.
CallPath/2 & CallPath DOS for Windows provide PS/2 Token Ring access to Northern Meridian 1 (X.25) and Rolm CallBridge via RS-422.
CallPath CICS provides AT&T ASAI/DCIU gateway to S/370/390 via SDLC and Switchserver/2 access to Northern Meridian 1 (X.25) and Rolm CallBridge via RS-422.
CallPath/400 interrface to AS/400 mini computer with 5250 terminal emulation.
Various incarnations of: Asynchronous, Token Ring, Ehternet, APPC, SNA (LU6.2, Netview) 3270; 5250 emulations.

IBM Interop Factor:

High - Besides DEC & AT&T, the best set of interop tools available.

IBM Alliances:

Rolm, Siemens, Intellivoice, Northern Telecom, Teleos, AT&T, Dialogic, InterVoice, Syntellect, ECMA (CSTA Tech. Comm. #32, TG11), ANSI (T1S1.1 Working Group), SCSA (Signal Computing System Architecture) supporter.

IBM Used In a variety of Systems:

Syntellect DT/2, Voicetek Generations, Logica C.O. Based Messaging, Aristacom SCIL*Link, Brock Activity Manager, Granada Systems Teldear, Datacorp Sales & Marketing Platform, Early Cloud Telephone Delivery System, InterVoice InterDial, NPRI Teltech Marketing, InteCom TimeClock

Call Processing Background:
IBM is a true veteran in call processing. Its Series 1 computer was the basis for one of the fist voice messaging systems called ADS (Audio Distribution System). It used proprietary cards in the S/1 backplane which connected to the telephone network. Series 1 computers were also the system of choice for one of the largest multi-city Audiotex networks installed by ACP (formerly Dial Info). The same platform was used in Telco-based operator services systems along with Votrax VRUs. Today, these operator services systems are IBM PC-based.

Direct Talk
The first incarnation of Direct Talk was a 68000-based VME VRU OEMd from Syntellect. Since then the PS/2 and RS/6000 have replaced that platform as the DirectTalk/2 and DirectTalk/6000, respectively. Ironically, Syntellect now re-sells IBM's software on IBM PS/2s along with Dialogic cards. These systems are packaged as DT/2 systems under a special arrangement with IBM.

The larger DirectTalk/6000 system was developed by IBM in LaGoude, France, and have an outboard "modem rack" type of box which houses call processing peripherals that are controlled from an RS/6000 host. AIX operating system drivers are available now for standard Dialogic cards, so the RS/6000 can also support call processing cards housed directly in its system bus as well. The proprietary "modem rack" configuration will probably go away in time, in favor of even higher density configurations available from vendors of SCbus cards.

CSA - CallPath Services Environment
Although DEC's CIT initiative is impressive in the area of CTI integrations, IBM seems to be on top when it comes to pull-through of its computer platforms. There are more business partners, VARs, resellers, telcos, and core technology suppliers using IBM connectivity tools (CSA) for datacom.

From a competitive standpoint, IBM has access to core technology, strategic liaisons, good connectivity (AT&T, Northern, Rolm, Siemens), and integrated platforms based on the PS/2 and RS/6000. All-in-all, they have their bases covered very well in this area and are ahead of the power curve from a market planning standpoint.

Harris DTS

"Harris believes that open systems standards for voice processing will provide system integrators with the opportunity to provide outstanding multi-vendor solutions to their customers. As a system integrator, we look forward to working with board suppliers to meet the worldwide needs of our customers." - Allen Jackson, Harris Digital Telephone Systems

Harris is another very credible addition to the list of SCSA supporters. The company has been involved in special enhanced services platforms for years, and its DTS group are expert in the area of systems integration.

The company has its own switch-to-computer link called HIL (Host Interface Link), which is used to control its 20/20 and Network Group Controller switches for custom applications. Harris is interested in working closely with Dialogic to define and refine specifications for host links to switches and also VRUs. This is where the ServerAPI technology comes into play. Harris certainly has a "leg up" in this area and has much to bring to the table in SCSA Working Groups and forums.

One of the company's newer products is a network-based platform called the Protocall 2000 Debit Card Calling System. The system allows callers to make prepaid calls from anywhere in the world using any touch-tone phone. By dialing an 800 access number, callers can access the platform, which reads their debit card number and validates it against a customer database in real time. This validation procedure allows calls to be placed, while the "balance" on the caller's pre-paid card is tallied "on the fly." The system provides automated prompting, operator escape, configurable warning messages, and call re-origination.

Figure **(HARRIS0.DRW)** illustrates the complexity of such a system. The incoming calls can be directed into the system via direct access or Feature Group access trunks. The Network Group Controller (NGC 2000) automatically connects callers over T-1 links to VRUs which prompt callers for their card number. The digits are sent down the Protocall System Bus to a validation unit. The VRU then signals back to the switch how to handle the call. Calls are then re-switched into the telephone network, handed-off to a live operator, or terminated.

HARRIS0.DRW

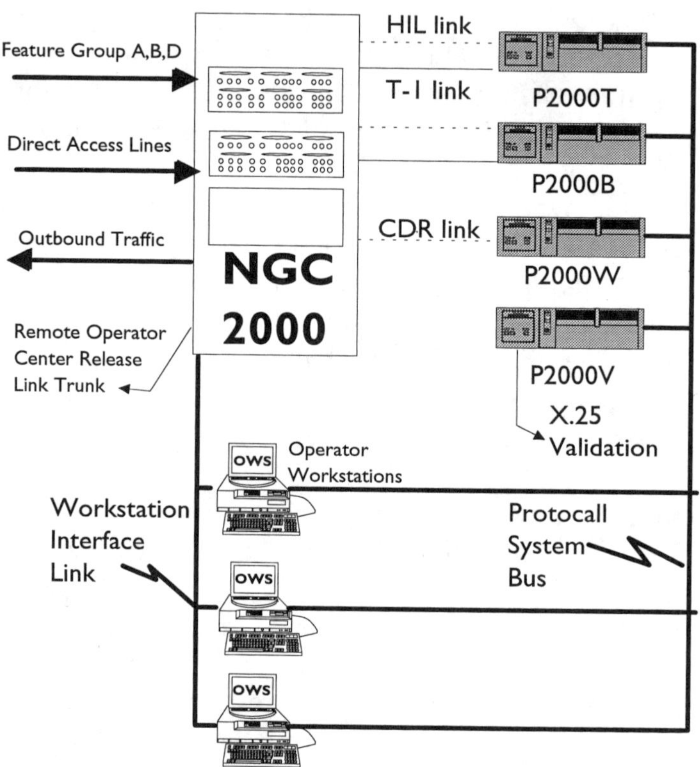

Harris Product Brief:

Model:	**Harris 20-20**	**Protocall 2000/NGC 2000**
Type Switch:	PBX/ACD	PBX/ACD
Resiliency Factor:	Mission Critical	Mission Critical
Simultaneous Timeslots:	960	960/node
Agent Queues:	32	32/node
Trunks:	1,920	1,920/node
seats/stations:	1,920/384	96/node
Price range:	$300	$450-1,000

Harris Typical Application Environments

Application	*Messaging*	*Audiotex*	*Fax S&F*	*Host IVR*	*Call Center*
CPE					☒
NTWK	☒			☒	☒

Harris Switch CPU:
Proprietary

Harris Operating System:
Proprietary

Harris Switch Backplane:
PCM/TDM

Harris CTI Interface:
Host Interface Link (HIL)
Asynchronous RS-232 (Specification Available)

Harris Station Gear:
Special Digital Display; Agent & Supervisor Stations based on PC/Windows, etc.

Harris Interop Factor:
Very high - Protocall/Voice Frame provides customized call center solutions including interactive PIN input, Audiotex, Host Interactive Response, and a robust Operator Services Package; Work Station Interface (WIL); Host Interface Link (HIL); SCSA (SCapi) development likely. ISDN PRI Support; European E-1 support; Switch can be configured as an access tandem for Feature Group A, B, D connectivity; release link trunks for remote node access.

Harris Alliances:
Cascade, Dialogic, SCSA endorser, Active Voice

Call Processing Background:
Harris is a solid player in enhanced services provision. Several years ago, the company acquired Protocall, an Operator Services systems integrator. Since then, the company has developed a suite of specialty services and capabilities around the 20-20, and have re-packaged a multi-node (hot-standby redundant) version for the high end called the Protocall 2000 Enhanced Services System. Along with its HIL link and Workstation Interface Link, the company can provide a turn-key, customized call center solution complete with integrated (Windows/GUI) agent terminals, host-coordinated call transfer, Audiotex, billing verification, and sophisticated host interactive services.

The Harris system also provides real-time routing of operator-assisted calls with outboard software state control. The systems can be multi-node and geographically distributed for nationwide call center capability. Harris products are also being used to upgrade the Federal Aviation Administration's air traffic control Voice Switching and Control System (VSCS), which requires a mission-critical grade of service.

NEC America, Inc.

> ***"By combining OAI architecture and SCSA, we can build applications that can grow and develop as customer needs evolve..."***
> ***- Dennis Soldner, NEC America, Inc.***

NEC has also developed a host-to-PBX link. In this case, it's called OAI Architecture, and is the basis for NEC's worldwide "Computers & Communications" thrust. The company has provided assistance to over 25 customers who use the special link over the past year, and about five of these have developed applications which company insiders think are commercially viable.

Dennis Soldner, a manager in product development, recently spoke to a group of SCSA supporters to explain NEC's position. He says that the company wants to help its customers to build application-specific communication systems, and that OAI and SCSA are a great way of doing that. According to Soldner, the OAI link can be used to control specific switch functions, and the ServerAPI link can be used to control VRU functions.

The company also is partial to its own proprietary phone functions on the NEAX switch, so a special phone emulation card is not outside of NEC's thinking. NEC wants to provide for the customization of standard communications systems, but also wants to extend the useful life of its switches. Although not obvious on the surface, this "life extension" philosophy is a new direction for NEC, who is known for fairly abrupt product terminations in the communications industry. Apparently, the company has learned some tricks from Northern and AT&T, who pioneered the "evergreen" concept for PBXs back in the late seventies.

Figure **(NEC0.DRW)** shows how NEC views SCSA, OAI, and Application-specific computing will come together in the call center. The company has yet to announce exactly what means will be used to connect the voice paths from the switch into the SCSA-based VRU.

NEC0.DRW

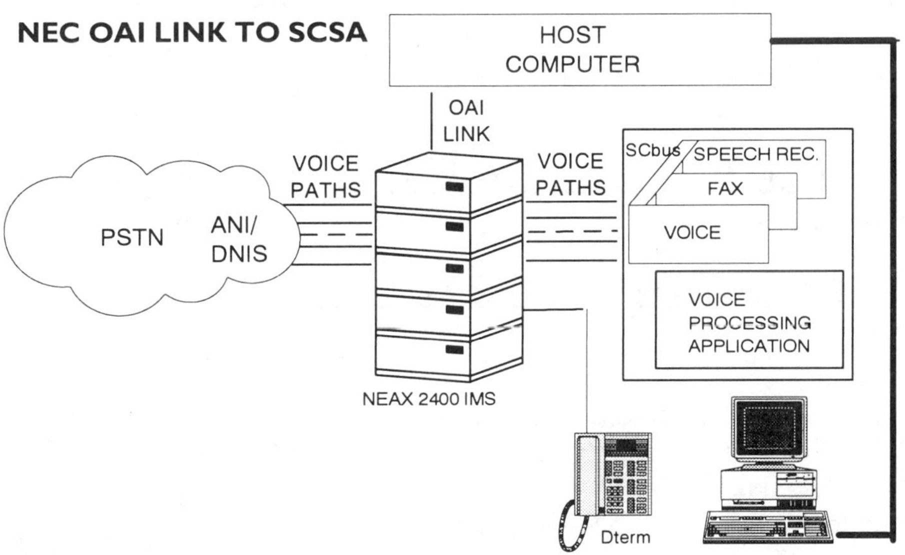

OAI Applications that are up and running include call center automation for health care, legal services, universities, and the hospitality industry. The company sees the marriage of OAI and SCSA as being helpful for call center applications, but wants to concentrate on the core switching capabilities of its existing line, so SCSA-based products are a "switched resource" off of the NEAX.

NEC Product Brief:

Model:	**NEAX 1400 IMS**	**NEAX 2400 IMS**
Type Switch:	PBX/ACD	PBX/ACD
Resiliency Factor:	Commercial	Commercial
Simultaneous Timeslots:	256	10,000+
Agent Queues:		25
Trunks:	256	3,000
stations/seats:	256	20,000
Price range:	$400	$350 - 550

NEC Typical Application Environments

Application	*Messaging*	*Audiotex*	*Fax S&F*	*Host IVR*	*Call Center*
CPE	☒			☒	☒
NTWK					

NEC Switch CPU:

Proprietary

NEC Operating System:

Proprietary

NEC Switch Backplane:

PCM/TDM

NEC CTI Interface:

NEAX 1400 - (none)
NEAX 2400 - NEC OAI
Proprietary and X.25 protocols supported on OAI Interface Processor (UNIX/SCO on 386/486 host). APIs are "C" routines.

NEC Station Gear:

NEAX 1400 - Dterm Series II Digital Display (proprietary LCD); POTS 2500.
NEAX 2400 - Dterm Series II Digital Display (proprietary LCD) X.25 data transfer between sets; POTS 2500.

NEC Interop Factor:

NEAX 2400 - Medium through proprietary
OAI link; PC Dterm emulation card under development.

NEC Alliances:

Supporter of SCSA architecture

Call Processing Background:

NEC launched its OAI initiative several years ago. OAI is NEC's as an answer to the "OPEN" issue and IBM's CallPath Services Environment spearheaded by IBM/ROLM/Siemens. To date, a handful of customers have co-developed vertical applications on the OAI link, which is a proprietary X.25 protocol. These call processing capabilities are a good complement to products such as the "Asset Master" This OAI program provides screen and database coordination for reserving conference rooms and meeting attendance; "Code Master" provides traveling class of service changes for authorized long distance - coordinated LCD prompts.

The company recently endorsed the SCSA architecture, and may be further opening its adjunct processor to a more open OAI link based on the SCapi command set. There are also initiatives underway to provide better interoperability with adjunct processors via digital set emulation on outboard PCs.

Northern Telecom

Northern Telecom locked horns with AT&T and ROLM in the early 80's over the "Voice/Data" Integrated switch. At that time, Northern launched its OPEN World Development Program, which sought to share switch integration specifications with vendors of computers, call accounting equipment, and voice mail systems.

Northern established technology sharing arrangements with DEC, Hewlett Packard, CommTerm, VoiceTek, Octel, and a host of other manufacturers. This initiative laid the foundation for what would become one of the most successful telephone switch products line in the world. Today, Northern Telecom is a leading supplier of not only key systems and PBXs, but also central office and tandem switching gear for the world's largest telephone companies.

The company boasts a full line of Interactive Voice Response and Messaging products for both its PBX and Key System Lines. Although they are based on different platforms, it is clear that the company has chosen the open architecture route. Northern is encouraging companies like Dialogic to produce software and hardware that will couple very closely to its switching products, especially for call center applications.

It's support of SCSA will most likely come in the form of support for SCbus-based hardware products that run in its new StarTalk II VRU. The StarTalk II product is slated for release within the year, and will offer PC-based voice processing and switch control through the Access Toolkit. The Access Toolkit is a DOS program which allows developers to control certain Norstar Switch functions.

In addition, the company is co-developing several PC cards which emulate the proprietary Norstar digital telephones. Figure **(NORSTAR.DRW)** is a cut-a-way view of a PC with Norstar emulation cards in the VRU. These cards will provide phone emulation functions as well as voice processing functions. Initially, they will be PEB bus compatible, but will most likely be upgraded to the SCbus in the near future.

NORSTAR.DRW

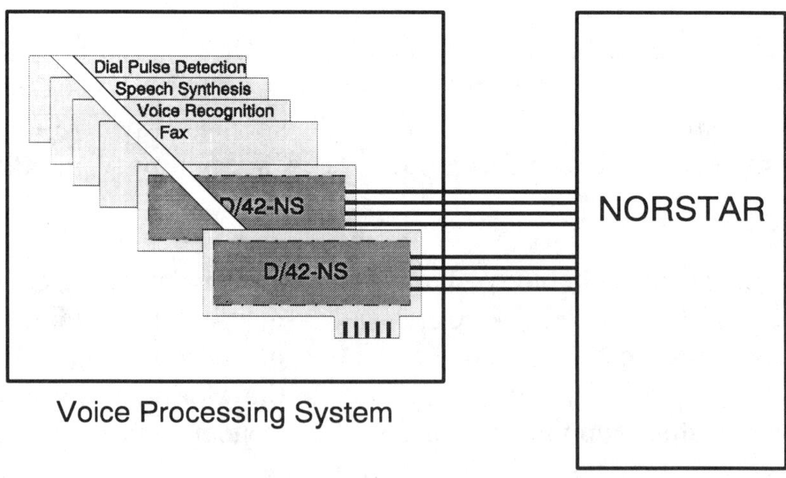

Northern is on the look-out for application developers who can bring call center, host interactive response, and facsimile applications to its Norstar customers. The company has enlisted Dialogic as a helper in its search for VARs who can do this. The basis for this search is the success of the 4-year old Norstar System (Key/Hybrid), which has met with great marketplace acceptance. The Norstar system has grown from 1% market penetration to about 17% since its introduction.

Northern Telecom Product Brief:

Model:	**Meridian Option 11**	**Meridian 1; ACD Max**	**Norstar**
Type Switch:	PBX/ACD	PBX/ACD	Hybrid/Key
Resiliency Factor:	Commercial	Redundant Option	Commercial
Simult. Timeslots:	256	10,000	256
Agent Queues:		100	
Trunks:	32	650+ ACD 4,000 PBX	144 (universal)
seats/stations:	30-224 stations	500 agents; 30-60,000	144 (universal)
Price range:	$650/line	$400+/line	$450/line

Northern Typical Application Environments

Application	***Messaging***	***Audiotex***	***Fax S&F***	***Host IVR***	***Call Center***
CPE	☒	☒	☒	☒	☒
NTWK				☒	☒

Northern Telecom Switch CPU:

Northern Proprietary.

Northern Telecom Operating System:

Northern SL (Proprietary)

Northern Telecom Switch Backplane:

PCM/TDM

Northern Telecom CTI Interface:

Meridian 1

Meridian Link; BRI ISDN; X.25; HDLC/LAPB (VAX)

Meridian Option 11

Meridian Telecenter (Mac/GUI Interface)

Norstar
Norstar Access/API & Toolkit (RS232 via PCIB card;
BRI/ISDN via DVC/PC Card or Dialogic D/42-NS PC Card).

Northern Telecom Station Gear:
Meridian; Digital Display; POTS phones; M300 Touchphone
Norstar; M7xxx series digital/LCD sets (BRI/NT-A10 chip set based)

Northern Telecom Interop Factor:
High; Meridain Link, SCAI (U.S. version of CSTA sponsored by ANSI) for Meridain; Norstar Access and Access Toolkit for Norstar, CompuCall X.25 or BRI links to adjunct processor for DMS-100 series of Central Office Switches.

Northern Telecom Alliances:
DEC CIT, HP ACT (3000/9000), IBM
CallPath (AS/400), Apple (Telecenter),
Tandem CAM (Non-Stop/CLX).

Call Processing Background:

Northern has a variety of solutions for integrated CTI available through NPRI, Datacorp, Brock Control Systems, and Dialogic VARs who integrate D/42-SL & D/42-NS digit display emulation interfaces. Both Ameritech & BellSouth have integrated Northern switches with DEC VAX computers.

With a 28-30% market share in PBX's, the company is a formidable competitor to AT&T, and provides roughly the same breadth of offerings to end users. The Meridian Mail and StarTalk voice messaging options sell into approximately 50% of all new installations or upgrades.

A variety of value-added resellers and computer companies are integrating NTI switches using CTI for Telemarketing, Operator Services, and Call Management applications. In addition to its own CTI interfaces, Northern recently endorsed the SCSA (Signal Computing System Architecture).

CHAPTER 12 - SCSA ToolKit Developers

No architecture is really complete without a suite of development tools. Fortunately, the most popular call processing development tools in the industry are developed by charter members of the Signal Computing System Architecture. Most of these companies are run by voice processing veterans, whose basic mission is to "kick-start" other companies into the voice and call processing business. They all provide industry newcomers with high-level languages and application generators which are compatible with industry-standard operating systems and components.

These call processing tool-makers will adapt their service creation products for new, higher density SCbus products, and will evolve their offerings to include client/server-based architectures and support for SCapi models and features. Since SCSA is a broad initiative, different tool providers will add value to the architecture in differing ways. For example, TRT (Telephone Response Technologies) has a head start on client/server architectures and large PC-based switches in their new Pro/Found and Pro/Switch products, respectively. Since they helped to pioneer these more sophisticated systems, they should be able to quickly adapt to higher-density SCbus products and APIs.

This chapter provides an overview of this and some of the most popular development tools available today. The companies represented are part of the OPEN ToolKit Development Program, which was formed about five years ago as a co-marketing venture between Dialogic and the industry's top tool providers. Literally thousands of industry newcomers are introduced to these development tools every year, and the products they manufacture are the backbone of both low and high-density systems worldwide:

- Telephone Response Technologies, Inc.
- Cascade Technologies, Inc.
- Expert Systems, Inc.
- Parity Software Development Corp.
- U.S. Telecom International, Inc.
- APEX Voice Communications, Inc.

Telephone Response Technologies (TRT)

"As products that support this new standard become available, a whole new class of dynamic voice processing applications will emerge. Systems will finally be possible that meet the demand for increased density and more features in the multi-application platforms of the future."
- Chris Bajorek, President

Telephone Response Technologies, Inc.
1624 Santa Clara Drive
Suite 200
Roseville, CA 95661
Phone: 916-784-7777
Fax: 916-784-7781
Contact: Chris Bajorek

TRT ProVIDE 4.2 Overview:

Operating System	Processor	Form Factor	Ports Supported
MS-DOS	Intel 386/486	ISA/MCA PC	2 - 48 lines
User Creation Interface	**Host Connectivity**	**LAN SUPPORT**	**Voice Recognition**
forms	yes	yes	yes

TRT Typical Application Environments

Application	*Messaging*	*Audiotex*	*Fax S&F*	*Host IVR*	*Call Center*
CPE	☒	☒	☒	☒	☒
NTWK		☒			☒

TRT Pricing Scheme:

ProVIDE 4.2 Applications Editor (AE) w/ 4-line Runtime AP- $650
AE & AP Starter Kit Including Voice Board and Audio Coupler - $1,295
ProVIDE/Link IBM 5250/AS400 - $4,995
ProVIDE/Link IBM 3278/3274 - $4,995
Pro/Found User Function & xBase Server Dev. License - $695
xBase Server Runtime (dBASE, Clipper, FoxPro, etc.) - $50 each
Client/Server Server Manager Dev. License - $995
Client/Server Server Manager Runtime - $495

TRT Alliances:
Dialogic, DiAnaTel, TeleVoice, Ipex, SCSA Supporter.
TRT Background:

TRT is a privately held corporation that has been in the application generator business for call processing solutions since the mid-eighties. TRT's ProVIDE package (formally called Intelesys) is easily the most popular of all of the MS-DOS based tools. The package is fully compatible with over a dozen Dialogic boards as well as switching and line interface products from DiAnaTel.

The ProVIDE offering is an integrated package of application development tools and runtime environment for producing and maintaining multiline voice and fax systems. ProVIDE is an IBM PC-compatible, MS-DOS-based package. The environment is a three-part system: An Applications Editor, Application Processor (runtime), and a Configuration Utility. The applications are developed in the Application Editor (AE), which presents the developer with a structured, forms-based interactive development environment. Both programmers and non-programmers can easily grasp the simplicity and straightforward nature of the editor. The AE's screen-based approach allows step-by-step creation of simple and very complex voice response applications.

The package includes an integrated suite of tools including an online interactive debugger and an extensive on-line hypertext help system. Some users claim the debugger alone cuts their development time by as much as 50%. As many as nine separate applications can be open at once inside of the AE, so developers can make common changes to programs very quickly.

The second part of ProVIDE is the Application Processor (AP), or runtime environment for ProVIDE applications. Up to 48 voice channels can be handled simultaneously per system node, with each line capable of handling its own application. Multiple systems can be placed on Novell LANs for even higher line densities, as with the Pro/Found client/server architecture which is covered below.

Advanced call status monitoring and reporting capabilities are included in the AP, and a variety of statistical counts can be logged to disk.

Customized usage reports can be generated on an hourly basis with the AP, which also handles sophisticated error handling. Error conditions can be reset while on-line without disturbing other lines or processes. Critical errors that cannot be recovered from automatically, such as out-of-disk-space problems, will immediately flag operator attention. Like the AE, the AP features a comprehensive hypertext system.

The final part of the three-part ProVIDE system is the Configuration Utility (CU). The CU is a menu-based interactive system management tool. Its menus and forms-based screens can be used to change and configure nearly every important runtime element of a ProVIDE system. In addition to the expansion modules, the three-part package includes built-in speech recognition hardware support and a module for supporting natural voicing of numerical, time, date, and dollar values called the SPEAK module.

An array of add-on software modules is available with ProVIDE:

INTELEFAX - Fax on Demand
TLOG - Transaction Logging Module
Pro/Link - Serial Comm. Link
User Function Toolkit
Pro/Switch - Resource Switching
Pro/Found Database Access Module

INTELECREDIT - Credit Card BVS
MMS - Message Management Module
P/Script - Procedural Language
Pro/Tone GTD, GTG, Speed, Volume
Database Servers (DB III, Clipper, etc.)

INTELEFAX - Fax Store & Forward

The INTELEFAX module adds "fax-on-demand" capabilities to ProVIDE. Callers can request specific data sheets, bulletins, and product guides and have them faxed-back instantly. Documents can be dynamically updated. The module is in use by several of the largest fax-on-demand vendors in the industry.

INTELECREDIT - Credit Card Billing Verification Service

The INTELECREDIT module adds automated credit card acceptance and verification capabilities to ProVIDE. Callers can enter a credit card number for order payment, and INTELECREDIT functions will validate the number. The module uses industry standard means of validation including the ***LUHN Mod-10*** algorithm used by banks and other financial institutions. A credit service can be automatically dialed for immediate authorization, or the transactions can be batched for after-hours processing.

TLOG - Transaction Logging Module

The TLOG module adds comprehensive caller transaction logging capabilities to ProVIDE. A customized call statistics report can be generated using this module much in the same manner as PBX SMDR (Station Message Detail Reporting) printouts are generated. Statistics including caller-entered data, line usage, phone numbers, and credit card numbers are typical with this module.

MMS - Message Management Module

The MMS adds a sophisticated database function which is the basis for message handling capabilities. By using MMS functions, ProVIDE applications can emulate standard voice messaging systems, answering services, and other devices that require the use of multiple message queues and password-protected objects.

ProVIDE/Link - Host Communications Link

Pro/Link adds host connectivity to ProVIDE applications. The Pro/Link module can be configured to handle between one and 32 external host communication links for sophisticated interactive response applications. For example, one Pro/Link port can be used to access information on a co-located computer's database, and another Pro/Link port can be used to remotely access a foreign computer simultaneously. The package has support for IBM mainframe and midrange terminals and controllers.

P/Script - Procedural Language

For those developers more accustomed to a high-level language and C-like interface, the P/Script option provides an alternative to the regular forms-based service creation environment of the AE. The P/Script is a powerful script programming tool that some developers prefer because of the lower-level flexibility of a procedural language. P/Script and ProVIDE code are interchangeable.

User Function Toolkit

The User Function Toolkit (UFTK) is a manifestation of the open architecture of the ProVIDE-based suite of expansion modules. It allows developers to create new software modules that bolt-on to TRT user functions. The toolkit provides reduced-memory functions that lower host overhead after compiling. UFTK is the same tool that TRT uses to produce new expansion modules.

Pro/Switch - Resource Switching

Pro/Switch is TRT's answer to the PC-based communications controller. Depending on the configuration and hardware used, Pro/Switch-based systems can emulate PBX functions, and also perform multi-span T-1 drop and insert functions for call center applications. Pro/Switch takes advantage of advanced hardware modules from Dialogic and DiAnaTel, including PRI ISDN line cards, AT&T Vari-A-Bill support, and full digital matrix switching. Pro/Switch is used by long distance carriers, operator services companies, and enhanced service providers.

Pro/Tone

The Pro/Tone module adds advanced DSP board features to ProVIDE. Full support of Dialogic's Global Tone Detection (GTD), Global Tone Generation (GTG), and speed and volume control are supported. The module is composed of a single terminate-and-stay-resident (TSR) program which allows developers to detect user-defined tones and tone pairs, given specific tone characteristics. In addition, Pro/Tone allows developers to generate user-defined tones or tone pairs from the application. By coupling these features with automatic gain and pitch-corrected speed control, virtually any analog PBX, hybrid key system, or central office interconnection can be achieved.

CLIENT/SERVER ARCHITECTURE and Pro/Found Module

Telephone Response Technologies has also developed a new Client/Server architecture which is consistent with the general direction of SCSA client/server models. The architecture will support a variety of collateral

products developed for ProVIDE, the first being a Database Access Module called Pro/Found.
The Pro/Found client/server module improves database access over LANs (Local Area Networks), and extends the growth of ProVIDE-based voice and fax applications. Systems formerly limited to several phone lines will now handle hundreds of simultaneous transactions. These larger systems are well suited to meet the needs of corporate communications, service bureaus, and call center environments.

The Client/Server extensions to ProVIDE signal a major contribution to the voice and call processing industry by providing developers access to LAN-based services, databases, and collateral technologies.

The Pro/Found module allows ProVIDE-based voice processing applications to access database files in dBASE III, dBASE IV, Clipper, FoxPro, FoxBASE, and Paradox formats. The Database Access Module enables this connectivity on both stand-alone PCs and LAN-connected client PCs with a Server Manager PC.

When systems utilize the Server, memory and processing requirements on the Client (voice) systems are significantly reduced, enabling greater line densities on those systems. This new Client/Server model enables DOS-based voice processing systems to realize the power of LAN connectivity and use it to great benefit when systems need to quickly expand.

TRT calls the new architecture **client/server** because of the decoupling of programs into separate elements; that is, a client part and a server part. Both the client program and the server program are executables that run as terminate-and-stay-resident programs (**TSR**s) on a call processing node. A **client** program is a User Function similar to TRT's other User Functions such as INTELEFAX, Pro/Switch, or TLOG. The first client program developed for this new client server architecture is Pro/Found. The difference between Pro/Found and these other User Functions is that Pro/Found relies on a second program, a database **server**, for interfacing with databases.

This is highly consistent with the direction of the ServerAPI specification as defined by the SCSA working group on APIs. The clear direction of the ServerAPI, and its first instantiation, called AppServer, is to evolve into a client/server paradigm. Currently, the ServerAPI only handles host/slave scenarios, but subsequent versions will have a similar look and feel to TRT's offering.

Pro/Found database servers are written for specific database formats. For example, the server program for dBASE III+ is called PF_DB3.EXE, the server for FoxPro is called PF_FOXP.EXE, and the server for Clipper is called PF_CLIP.EXE.

The client/server architecture implemented by Pro/Found offers three unique advantages to application developers:

- Applications can be written without explicit knowledge of the actual database format which will be used.
- The database format can be changed (from Clipper to FoxPro, for example, or dBASE IV to Paradox) with no changes to the application.
- The client and server programs can be run on separate, network-linked computers.

Because the Pro/Found Do/Change functions do not make assumptions about the underlying format of a database or its files, it is possible to write a single application which can use databases in the dBASE, Clipper, FoxBASE, FoxPro or Paradox formats. This database independence allows application developers to distribute the application to a customer who wants to use existing dBASE files, and the same application to another customer who wishes to use Clipper data files.

Local vs. Distributed Modes

The client/server architecture can be used with both modules running on the same PC (local mode) or with the server module running on a separate, network-connected system (distributed mode).

Client/Server and Pro/Found Elements on a single-node system:

Client/Server Element	Function
Applications Processor	Runtime Program
Pro/Found User Function	Client-Side User Function TSR
Database Server	xBase Driver (DB III, Clipper, etc.)

Client/Server and Pro/Found Elements on a distributed multi-node system:

Client/Server Element	Function
Applications Processor	Runtime Program
Pro/Found User Function	Client-Side User Function TSR
Client Manager	Client-Side Inter-Node Communications
Server Manager	Server-Side Inter-Node Communications
Database Server	xBase Driver (DB III, Clipper, etc.)

Many Pro/Found installations will consist of a single computer running the Application Processor (the runtime application program), and PROFOUND.EXE (the client-side user function), and a database server program such as PF_CLIP.EXE. When both the client and the server are running on the same computer Pro/Found is said to be operating in local mode. In environments such as this, Pro/Found executes Do/Change functions and acts entirely as if the PROFOUND.EXE and the server program were a single program. This type of installation is shown in **Figure (PFOUND1.DRW).**

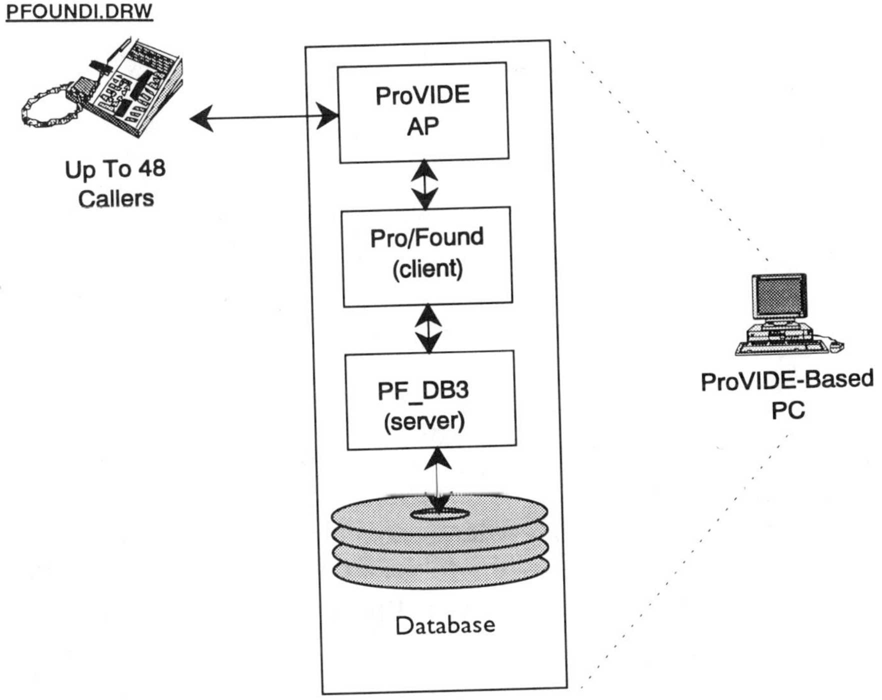

Distributed Mode

Some applications will benefit from operating Pro/Found in **distributed mode**, which means that the client and server programs are running on two different computers. The computers must be connected either through a Novell network or through any **NetBIOS**-compatible network. A typical distributed-mode system is shown in **Figure (PFOUND2.DRW).**

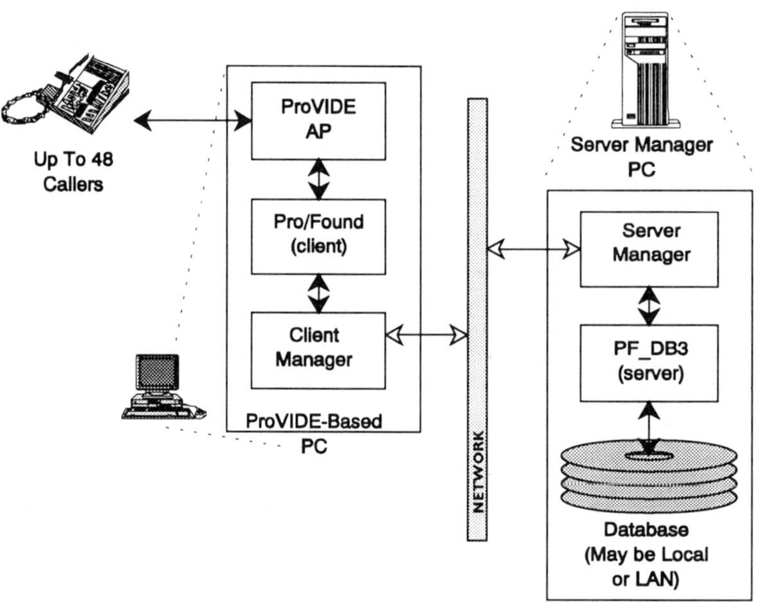

Applications can be written for use in local mode and then used in distributed mode *without a single change to the application code.* In fact, because of Pro/Found's approach of database independence, applications can be written for local mode use with, for example, dBASE IV and then used in distributed mode with Paradox again with no application code changes.

Some of the major advantages to running Pro/Found in distributed mode include:

- Memory requirements on the "run-time" system are significantly reduced.

- Database processing overhead is moved to the server machine, reducing its impact on the "run-time system's" performance.

- Multiple run-time systems can make use of the same server system and database.

In distributed mode Pro/Found goes beyond connecting a single **client system** (i.e., the machine running the Application Processor Runtime and the Pro/Found executable to a single server system (i.e., the computer running a database server such as PF_DB4.EXE). It is possible to create installations such as that shown in **Figure (PFOUND3.DRW)**, with two or more client machines using the same database server.

Note that although the illustrations show the database as being part of the server system, the database files can actually be located anywhere on the network file path. Thus the voice system could be located on one node, the database server program on another, and the database itself on the network file server.

The chief advantage to using multiple client systems with a single server system is that many applications require database access, but not enough of it to justify full-time use of a database server. The client/server architecture allows AP (run-time) nodes on a network to share the database server as a resource. Just as one company librarian can assist many employees with information requests, a database server can handle the needs of many client applications.

PFOUND3.DRW

Up To 48
Callers
ProVIDE
AP
Pro/Found
(client)
Client
Manager
ProVIDE-Based
PC
Up To 48
Callers
ProVIDE
AP
Pro/Found
(client)
Client
Manager
NETWORK
Server Manager
PC
Server
Manager
PF_DB3
(server)
Database
(May be Local
or LAN)

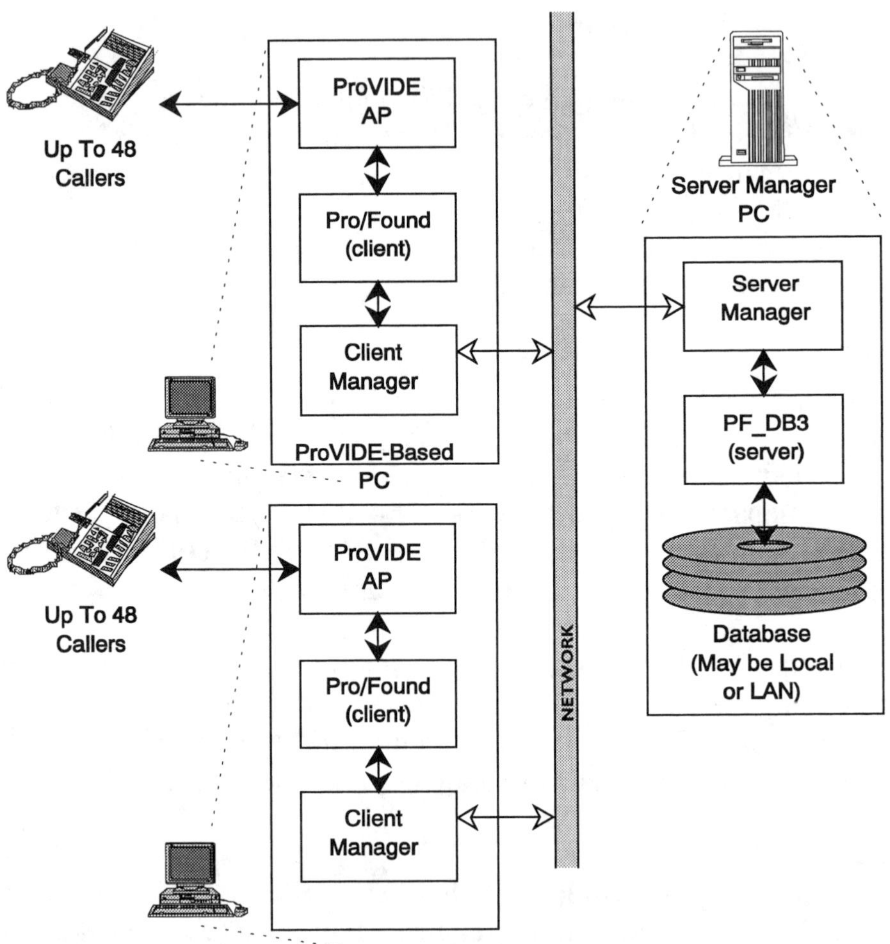

Cascade Technologies, Inc.

"SCSA is the right combination of advanced call processing hardware and object-oriented software for meeting both developer and customer demands." - Frank Joicy, Vice President.

Cascade Technologies, Inc.
1430 Broadway
New York, New York 10018
Phone: 212-768-7380
Fax: 212-768-7806
Contact: Neil Swanson

Cascade Voice Tools Product Overview:

Operating System	**Processor**	**Form Factor**	**Ports Supported**
OS/2	Intel 386/486	ISA/MCA	2 - 24 lines
User Creation Interface	**Host Connectivity**	**Speech Synthesis**	**Voice Recognition**
Forms	Yes	no	yes

Cascade Typical Application Environments

Application	***Messaging***	***Audiotex***	***Fax S&F***	***Host IVR***	***Call Center***
CPE			☒	☒	☒
NTWK					

Cascade Pricing Scheme:

Development Environment

VTAG4 - $1,995
VTAG8 - $2,495
VTAG12 - $2,995
VTAG16 - $3,395
VTAG24 - $3,995

VoiceTools RunTime

VTRT4 - $995
VTRT8 - $1,195
VTRT12 - $1,395
VTRT24 - $1,795

Cascade Alliances:
Dialogic, Harris DTS, SCSA Supporter

Cascade Background:

Cascade Technologies is one of a handful of call processing tool developers who support OS/2. The company's expertise is based in benefits platforms and host interactive response. The current Cascade Voice Tools package was originally dubbed "CAS Voice," which stands for Capital Accumulation System.

Fortune 500 companies use the turn-key system to provide telephone access to employees who want to inquire as to the balance of certain benefits including retirement, profit sharing, and stock options. Cascade has also developed a variety of fax response extensions to its product line, and provides (Excel-like) commands to calculate complicated formulas including mortgage rates and sale prices.

Cascade Voice Tools is a high-level, easy-to-use applications generator for creating multiline voice processing platforms. With Voice tools, developers with limited programming experience can create and maintain interactive voice response systems. The package provides the application developer with a series of step-by-step visual menus that represent calling options normally presented to someone dialing into the system. This visual "dial pad" layout allows programmers to create applications intuitively, claims Cascade. By mimicking the caller's options, the Voice Tools graphical interface keeps the application developer focused on the end-user calling experience.

The Voice tools development environment guides the developer through the application design process. Voice menus and responses appear on-screen, so the developer can easily design and maintain an application that is simple to use. Data references, calculations, and conditional references can be created and displayed without any previous programming knowledge. Cascade has a Toolbox that allows for custom "C" routines as well as simple ASCII and dBASE III file access methods.

Cascade also provides a library of professionally recorded phrases including number, day, and date information in several formats. The

Voice Tools package can use voice files created by DOS-based or Windows packages including the popular VFEdit, by Voice Information Systems, Inc.

The Voice Tools package is designed for either dedicated use by a single client or for shared use by multiple clients in a service bureau environment. The package also generates reporting data for billing purposes.

Expert Systems, Inc.

"The future of call processing lives at the intersection of computing and telephony. SCSA enables this future." - Carlton Carden, President

Expert Systems, Inc.
1301 Hightower Trail
Suite 201
Atlanta, GA 30350
Phone: 404-662-7575
Fax: 404-587-5547
Contact: Phil Hoffman

Expert Systems EASE Product Overview:

Operating System	Processor	Form Factor	Ports Supported
MS-DOS	Intel 386/486	ISA/PC	2 - 48 lines
User Creation Interface	**Host Connectivity**	**Speech Synthesis**	**Voice Recognition**
Forms & Procedural	yes	no	no

Expert Systems Typical Application Environments

Application	*Messaging*	*Audiotex*	*Fax S&F*	*Host IVR*	*Call Center*
CPE		☒		☒	
NTWK		☒			

Expert Systems Pricing Scheme:

EASE DOS Development License - $595 - 5K
EASE DOS Run Time License - $395 - 3K

SPL DOS license - $2.5 - 5.5K
xBase Development License (dBASE, Clipper, etc.) - $695
xBase Runtime License - $345

Host Computer Interfaces:
2 - 8 port RS-232 Asynch - $275 - 525
IBM 3174/3274 Emulation - $550 - 1K
IBM 3278 Terminal Emulation - $550 - 1K
IBM 5251 Emulation - $550 - 1K

Expert Systems Alliances:
Dialogic, Cognitronics, Microlog, DiAnaTel, SCSA supporter.

Expert Systems Background:

Expert Systems is a call processing company with host interactive response experience. Formed by former Cox Cable employees, the company has developed a variety of voice processing platforms before standardizing on Dialogic for access to core technology.

The company has sold turn-key home banking systems, cable pay-per-view, and benefits management platforms. Expert has ported its DOS environment to UNIX for Cognitironics, who Expert refers UNIX business to. In addition, Expert has OEM deals with several traditional voice mail manufacturers for ancilliary host connectivity products.

For the mainstream DOS-based EASE product, Expert houses the voice processing and line interface cards in a PC. EASE itself is a PC AT-compatible applications generator that allows both programmers and non programmers to develop applications quickly. The EASE environment is a menu-driven, pop-up window development system that uses Dialogic hardware.

The company claims to support as many as 96 lines and a wide range of host computer interfaces. Like other tool providers, Expert Systems offers both a high-level development environment, a script language, and C-function hooks. The procedural language is called SPL, or Scenario Programming Language.

EASE provides in-memory, disk-based, and NOVELL LAN-based database access as well as communication links to foreign databases. The package includes single-port RS-232 support for asynchronous communication to host such as DEC, Tandem, and PCs, for example. Expert also offers several cluster controller and terminal emulation options.

Pre-set default parameters are used in the EASE package to shorten development cycles. Although an on-line debugger is not available, the system does offer automatic error, timeout, and disconnect handling like other application generators. These features significantly reduce redundant code generation. System diagnostics, and integrated data scope, and an Error correction facility can quickly help programmers to pinpoint bugs. EASE also generates reports which can document a program and its associated logic flow after the application is written.

Parity Software Development Corp.

"Parity Software believes that the SCSA standard will greatly enhance the ability of OEMs and VARs to deliver cost-effective voice processing solutions..." Bob Edgar, President

Parity Software Development
25 Stillman Street, Suite 106
San Francisco, CA 94107
Phone: 415-931-8221
Fax: 415-546-7329
Contact: Bob Edgar

Parity VOS Product Overview:

Operating System	**Processor**	**Form Factor**	**Ports Supported**
MS-DOS	Intel 386/486	ISA/PC	2 - 48 lines
User Creation Interface	**Host Connectivity**	**Speech Synthesis**	**Voice Recognition**
textual, procedural	yes	yes	yes

Parity Typical Application Environments

Application	*Messaging*	*Audiotex*	*Fax S&F*	*Host IVR*	*Call Center*
CPE		☒	☒		☒
NTWK		☒			

Parity Pricing Scheme:

Development license for VOS - $1,885
C Developer's Toolkit - $1,295
Voice Recognition Toolkit - $1,295
Fax Developer's Toolkit - $1,295

Parity Alliances:

Systems Integrators, SCSA supporter

Parity Background:

Parity Software Development provides a script-like language for building call processing systems in DOS. The company also provides optional modules for the development of applications that require core technologies other than voice. Most of Dialogic Corporation's voice, fax, and voice recognition technologies are supported by Parity. The company has also developed host communication links with IBM mainframes and minicomputers, and VOS supports several databases including dBase and Btrieve.

The company claims that its products are simple enough that any programmer with a few years of experience with BASIC code will be producing VOS-based code right away: The language uses named variables and arrays, arithmetic expressions, Boolean logic commands, and functions that have named parameters.

Although the VOS language is fairly robust, it is not the kind of product that a non-programmer can tackle. It's look and feel is just slightly higher than the C library routines offered for free by board makers such as Dialogic and Rhetorex. In the example that follows, VOS source code defines a new command "menu" to play a prompts and get a single touch-tone response in the range 1... max. Up to three attempts are made to get a valid digit in this example:

```
func menu(prompt, max)

        dec
                var try  :  1;
                var digit  :  1;
        endec

# Loop to make up to three tries

        try = 1;
        while (try<=3)

# Play prompt message


        msg_play(prompt) ;

# Get 1 digit from caller

                digit = gtdgts(1)
;

# Check digit in wanted range

                if (digit>=1
and digit<=max)
                        return
digit ;
                endif
                try = try + 1;
        endwhile

# 3 tries failed, play message,
# return zero to indicate failure

        msg_play(ERROR) ;
        return 0;
endfunc
```

U.S. Telecom International, Inc.

U.S. Telecom International
211 Main Street
Suite 401
Joplin. MO 64801
Phone:417-781-7000
Fax: 417-623-2963
Contact: David King

U.S. Telecom VAL Product Overview:

Operating System	**Processor**	**Form Factor**	**Ports Supported**
MS-DOS	Intel 386/486	ISA/PC	2 - 64 lines
User Creation Interface	**Host Connectivity**	**Speech Synthesis**	**Voice Recognition**
Textual, Procedural	no	no	yes

U.S. Telecom Typical Application Environments

Application	***Messaging***	***Audiotex***	***Fax S&F***	***Host IVR***	***Call Center***
CPE	☒	☒			
NTWK		☒			

U.S. Telecom Pricing Scheme:

Licensed on a per-line run-time basis.

U.S. Telecom Alliances:

Education, Local Government, Dialogic, SCSA Supporter.

U.S. Telecom Background:

U.S. Telecom is a Joplin, Mo. based company that specializes in application generation tools and also has a turn-key business for outbound dialing and notification systems. The notification systems are sold to school systems and provide homework hotlines and truancy reporting.

VAL is a segmented service creation and run time environment for MS-DOS systems which provides rapid development tools for people who are familiar with procedural programming. The run time module provides the capability to "partition" as many as 65 processes, which are usually related to voice ports or other devices such as serial I/O or other application programs compiled under DOS.

With VAL, programmers can use any text editor to write programs that read like verbal descriptions of an application's flow. The programming model is based on a single-threaded, fall-through logic. U.S. Telecom says that VAL can be used to create both voice and non-voice programs, because it is a full programming language. For example, the VAL runtime can support additional "pseudo-ports" on which non-voice applications such as database, communications, and other functions can run.

The programming model allows separate applications to be "chained," or allocated to any port on the fly. For example, a caller may enter a code or speak a command that causes the program to switch from a voice messaging application over to a host interactive session.

In addition to support of Dialogic's switching, line interface, and voice store and forward technologies, VAL can also handle voice recognition capabilities on the VR/121 and VR/40.

Apex Voice Communications

APEX Voice Communications
14900 Ventura Blvd.
Suite 200
Sherman Oaks, CA 91403
Phone: 818-379-8400
Fax: 818-379-8410
Contact: Ben Levy

"SCSA is the next intelligent step for call processing. By integrating features such as increased bus bandwidth and third-party compatibility along with existing standards, SCSA puts call processing in the main stream of business solutions." - Ben Levy, President

Apex OmniVox Product Overview:

Operating System	**Processor**	**Form Factor**	**Ports Supported**
UNIX V	Intel 386/486	ISA Passive Backplane	up to 100+ using NFS networking
User Creation Interface	**Host Connectivity**	**Speech Synthesis**	**Voice Recognition**
Character-Based, Forms	SNA/3270;5250 X.25; RS-232; Ethernet	no	yes

Apex Typical Application Environments

Application	***Messaging***	***Audiotex***	***Fax S&F***	***Host IVR***	***Call Center***
CPE	☒	☒	☒	☒	☒
NTWK		☒			☒

Apex Pricing Scheme:

4-port license - $3,000
24 - port license - $24,000

Apex Alliances:

Service Bureaus, Foreign PTTs, Dialogic, SCSA Supporter

Apex Background:

Apex was established in 1987 and currently has 12 employees. Apex is a fairly diverse company with extensive international experience. The company has a broad range of core technology support including E-1, T-1, PRI ISDN, Voice Recognition, and Fax.

The bulk of Apex' revenues stem from the sale of turn-key, custom messaging and high-end Audiotex systems. The OmniVox applications generator comes with a variety of options including customer name and address capture (CNA), voice messaging capability, and fax store and forward. Large virtual system sizes can be achieved by networking multiple VRUs together. VRUs are typically one to two T-spans (24-48 lines) capacity. C language development tools are available for hard-coding underneath the OmniVox application generator for custom application developers.

OmniVox contains several integrated application development tools. The basic element of OmniVox is the Voice Applications Generator. Along with basic voice recording and playback capabilities including touch-tone call control, the Voice Applications Generator handles general information processing including data entry and retrieval, data conversion, and reporting. In addition, as with most application generators, there are facilities available for hard-coding user functions in C.

The Voice Applications Generator segments call logic flow into basic functions which are represented on a forms-based series of screens. Some very complex applications can be built using these forms, including the capability to handle Dialed Number Identification Service (DNIS), Outdialing, and live transfers.

Appendix A - Additional Resources

The Guide to T-1 Networking - by Bill Flanagan, 1990

Newton's Telecom Dictionary - by Harry Newton, 1993

Frames, Packets & Cells for Broadband - by Bill Flanagan, 1991

The Guide To Frame Relay - by N. Muller & R. Davidson, 1991

Local & Long Distance Billing Practices - by M. Brosnan & J. Messina, 1993

Understanding Data Communications - G. Friend & J. Fike, 1993

Understanding Fiber Optics - by Jeff Hecht, 1993

Understanding Telephone Electronics -by J. Fike, 1993

Handbook of Telecommunications - by James Harry Green, 1993

Telecommunications Management- by James Harry Green, 1993

Basic Book of Information Networking - by Motorola Codex, 1992

Basic Book of ISDN - by Motorola Codex, 1992

Basic Book of OSI & Network Mgn't - by Motorola Codex, 1992

Basic Book of Frame Relay- by Motorola Codex, 1992

Basic Book of x.25 Packet Switching - by Motorola Codex, 1992

The Guide to Sonet - by R. Davidson & Nathan Muller

Complete Traffic Engineering Handbook - Jerry Harder, 1992

Speech Recognition - Pete Foster, 1993

These titles avaliable from
TELECOM LIBRARY INC.
12 WEST 21ST ST
NEW YORK, N.Y. 10010
212-691-8215
1-800-LIBRARY

API Definition and Specification for thE *R 1.0 Netware* Telephony Server, Preliminary Draft Issue 1.3, March 31, 1993. AT&T Global Business Communications and Novell Inc.

Guide To Getting Started Profitably, Version 2.0, October 1993. Telephone Response Technologies, Inc.

Microsoft At Work Architecture Backgrounder, June 1993. Microsoft Corporation

Signal Computer System Architecture: Specification, March 1993. Dialogic Corporation, part number 05-0504-003

Telephone Manager Developer's Kit, Version 1.0. Apple Computer, Inc.

White Paper: The IAD - Online Interactive Debugger (For Voice Processing Development), Version 1.0, August 1993. Telephone Response Technologies, Inc.

Windows Telephony API (Preliminary), May 3, 1993. Microsoft Corporation.

Special Edition Of The Voice Processing Glossary for SCSA: The Book

Fourth, Updated Edition including SCSA / Client-Server Telecomputing Terms

Prepared by the research staff of TELECONNECT and CALL CENTER Magazines and Telecom Library Inc. under the direction of Harry Newton.

A & B BITS
Bits used in digital environments to convey signaling information. A bit value of one generally corresponds to loop current flowing in an analog environment. A bit zero corresponds to no loop current, i.e. to no connection. Other signals are made by changing bit values. For example, a flash-hook is set by briefly setting the A bit to zero.

A & B LEADS
Additional leads used typically with a channel bank two-wire E&M interface to connect certain types of PBXs (also used to return talk battery to the PBX).

A & B SIGNALING

Procedure used in most T-1 transmission links where one bit, robbed from each of the 24 subchannels in every sixth frame, is used for carrying dialing and control information. A type of in-band signaling used in T-1 transmission, A and B signaling reduces the available user bandwidth from 1.544 to 1.536 Mbps. See also PRI.

A LAW

The PCM coding and companding standard used in Europe. See A LAW ENCODING and E-1.

A LAW ENCODING

The method of encoding sampled audio waveforms used in the 2048K bit 30 channel PCM primary system, widely used outside North America. See E-1.

ACCESS TANDEM

A special type of local phone company central office. It's designed to provide equal access for all the long distance carriers in that area.

A/D

Analog to digital conversion.

ADAPTIVE PULSE CODE MODULATION

A way of encoding analog voice signals into digital signals by adaptively predicting future encodings by looking at the immediate past. The adaptive part of this method reduces the number of bits per second that another rival and more common method called PCM (Pulse Code Modulation) requires to encode voice. Adaptive PCM is not especially common because, even though it reduces the number of bits required to encode voice, the electronics to do it are expensive. See also PULSE CODE MODULATION.

ADC
Analog-to-Digital Converter.

ADPCM
Adaptive Differential Pulse Code Modulation. A speech coding method which calculates the difference between two consecutive speech samples in standard PCM coded voice signals. This calculation is encoded using as adaptive filter and therefore, is transmitted at a lower rate than the standard 64 Kbps technique. Typically, ADPCM allows an analog voice conversation to be carried within a 32k-kbit/s digital channel; 3 or 4 bits are used to describe each sample, which represents the difference between two adjacent samples. Sampling is done 8,000 times a second.

ADSI
Analog Display Services Interface. ADSI is a Bellcore standard defining a protocol on the flow of information between something (a switch, a server, a voice mail system, a service bureau) and a subscriber's telephone, PC, data terminal or other communicating device with a screen. The simple idea of ADSI is to add words to, and therefore a modicum of simplicity of use to a system that usually uses only touchtones. Imagine a normal voice mail system. You call it. It answers with a voice menu. Push 1 to listen to your messages, 2 to erase them, 3 to store them, 4 to forward them, etc. It's confusing. You have to remember which is which. ADSI is designed to solve that. It's designed to send to your phone's screen the choices in words that you're hearing. You then have the choice of reponding to what you hear or what you see. Your response is the same -- a touchtone button. ADSI's signaling is DTMF and standard Bell 202 modem signals from the service to your 202-modem equipped phone. From the phone to the service it's only touchtone. With ADSI, you don't hear the modem signaling because every time the service gets ready to send you information, it first sends a "mute" tone. ADSI works on every phone line in the world.

ADVISORY TONES
Signals such as dial tone, busy, ringing, fast-busy, call-waiting, camp-on-- and all the other tones which your telephone system uses to tell you that something is happening or about to happen.

AEB
Analog Expansion Bus. Dialogic's name for the analog electrical connection between its network interface modules and its analog resource modules. This bus is now "open." Technical specifications are available, thus enabling outsiders to make their own resource modules and/or network interface modules. The AEB interfaces DTI/124 and D/4x voice response component boards which fit in an AT-expansion slot of a PC. See also PEB, which is the more modern digital PCM expansion bus. See also SCSA, which is Dialogic's latest standard.

AGC
Automatic Gain Control. There are two electronic ways you can control the recording of something -- Manual or Automatic Gain Control (AGC). AGC is an electronic circuit in tape recorders, speakerphones, and other voice devices which is used to maintain volume. AGC is not always a brilliant idea, since AGC will attempt to produce a constant volume level. This means it will try to equalize all sounds -- the volume of your voice, and, when you stop talking, the circuit static and/or general room noise which you undoubtedly do not want amplified. Sometimes it's better to have quiet, when you want quiet. Manual Gain Control is preferred in professional applications.

ALERTING SIGNAL
A ringing signal put on subscriber access lines to indicate there's an incoming call.

ALGORITHM
A prescribed finite set of well defined rules or processes for the solution of a problem in a finite number of steps. In normal English, it is the mathematical formula for a computing operation.

ALL TRUNKS BUSY
When a user tries to make an outside call through a telephone system and receives a "fast" busy signal (twice as many signals as a normal busy in the same amount of time), he is usually experiencing the joy of All Trunks Busy. No trunks are available to handle that call. The trunks are all being used at that time for other calls or are out of service. These days, many long distance companies are replacing a "fast" busy signal with a recording that might say something like, "I'm sorry. All circuits are busy. Please try your call later."

AMIS
See AUDIO MESSAGING INTERCHANGE SPECIFICATION. AMIS is a standard for networking voice mail systems.

AMPLITUDE
The distance between high or low points of a waveform or signal. Also referred to as the wave "height."

AMPLITUDE DISTORTION
The difference between the output wave shape and the input wave shape.

AMPLITUDE EQUALIZER
A corrective network that is designed to modify the amplitude characteristics of a circuit or system over a desired frequency range. Such devices may be fixed, manually adjustable, or automatic.

AMPLITUDE MODULATION
Also called AM, it's a method of adding information to an electronic signal in which the signal is varied by its height to impose information on it. "Modulation" is the term given to imposing information on an electrical signal. The information being carried causes the amplitude (height of the sine wave) to vary. In the case of LANs, the change in the signal is registered by the receiving device as a 1 or a 0. A combination of these conveys different information, such as words, numbers or punctuation marks.

AMX
Analog Matrix Switch. A Dialogic product. An 8 x 8 analog crosspoint switch on a board used for sharing resources among AEB-based products and for connecting to phones and external audio devices.

ANALOG
Comes from the word "analogous," which means "similar to." In telephone transmission, the signal being transmitted -- voice, video, or image -- is "analogous" to the original signal. In other words, if you speak into a microphone and see your voice on an oscilloscope and you take the same voice as it is transmitted on the phone line and ran that signal into the oscilloscope, the two signals would look essentially the same. The only difference is that the electrically transmitted signal (the one over the phone line) is at a higher frequency. In correct English usage, "analog" is meaningless as a word by itself. But in telecommunications, analog means telephone transmission and/or switching which is not digital. See ANALOG TRANSMISSION.

ANALOG BRIDGE
A circuit which allows a normal two-person voice conversation to be extended to include a third person -- without degrading the quality of the call. Digital bridges work better.

ANALOG/DIGITAL CONVERTER
An A/D Converter. Pronounced: "A to D Converter." A device which converts an analog signal to a digital signal.

ANALOG FACSIMILE
Facsimile which can transmit and receive grey shadings -- not just black and white. "Analog" facsimile is usually transmitted digitally. But it's called analog because of its ability to transmit what appear to be continuous shades of grey.

ANALOG TRANSMISSION
A way of sending signals -- voice, video, data -- in which the transmitted signal is analogous to the original signal. In other words, if you spoke into a microphone and saw your voice on an oscilloscope and you took the same voice as it was transmitted on the phone line and threw that signal onto the oscilloscope, the two signals would look essentially the same. The only difference between them is that the electrically transmitted signal would be at a higher frequency.

ANI
Automatic Number Identification. A phone call arrives at your home or office. At the front of the phone call is a series of digits which tell you the phone number calling you. These digits may arrive in analog or digital form. They may arrive as touchtone digits inside the phone call. They may arrive in a digital form on a separate circuit. Whichever way they arrive, you will need some equipment to decipher the digits AND do "something" with them. That "something" might be throwing them in a database and bringing up your customer's record on a screen in front of your telephone agent as he answers the phone. "Good morning, Mr. Smith." Some large users say they could save as much as 20 seconds on the average IN-WATS call if they knew early on the phone number of the person calling them. They wouldn't need to ask their regular customers for their address, phone number, credit card number, etc.

ANNOUNCEMENT SERVICE
Allows a phone user to hear a recording when he dials a certain number. Could be an answering machine. Could be a voice response system. Could also be a solid-state digital announcer. Increasingly, it's a voice response system.

ANSWER DETECT
The use of a digital signal processing technique to determine the presence of voice energy on a telephone line. It is used with call (answer) supervision, to identify an answered line. It is beginning to be used with computerized dialing equipment in business to consumer calling. This technique eliminates the need for a telephone representative to constantly monitor call set-up progress on each telephone line in the event a call is answered. See ANSWER SUPERVISION.

ANSWER SUPERVISION
Follow this scenario: I call you long distance. My central office must know when you answer your phone, so my central office or long distance phone company can start billing me for the call. It works like this: when you, the called party, answer your phone, your central office sends a signal back to my central office (the originating CO). This tells my central office to start billing me for the call. This signal is called Answer Supervision. Before the Divestiture of the Bell System in early 1984, most of the nation's long distance companies -- with the exception of AT&T -- did not receive Answer Supervision. They did not know precisely when the called party answered. So they started their billing cycle after some time -- 20 or 30 seconds after the caller completed dialing. These long distance companies presumed that after this time, some one will have answered and the call will be in progress. Therefore start timing (and billing) for the call. Without Answer Supervision, their billing of calls is inaccurate. With the Divestiture of the Bell System, and the introduction of Equal Access, most long distance companies now receive true Answer Supervision. See also SOFTWARE SUPERVISION.

ANTI ALIASING
A computer imaging term. A blending effect that smooths sharp contrasts between two regions of different colors. Properly done, this eliminates the jagged edges of text or colored objects. Used in voice processing, anti-aliasing usually refers to the process of removing spurious frequencies from waveforms produced by converting digital signals back to analog.

API
Application Programming Interface. "Hooks" into software. A set of standard software interrupts, calls, and data formats that application programs use to initiate contact with network services, mainframe communications programs, or other program-to-program communications. For example, applications use APIs to call services that transport data across a network. Standardization of APIs at various layers of a communications protocol stack provides a uniform way to write applications. NetBIOS is an early example of a network API. Also, applications use APIs to call services that transport data across a network. See also WINDOWS TELEPHONY.

APPLICATION
A software program that carries out some useful task. Database managers, spreadsheets, communications packages, graphics programs, and word processors are all applications.

APPLICATION PROGRAM INTERFACE
A set of formalized software calls and routines that can be referenced by an application program to access underlying network services.

APPLICATIONS GENERATOR
Basically, software that writes software. Applications generators are software tools that, in response to your input, write software code which a computer can understand. Applications generators have two major benefits: First, they save time. An applications generator is perfect for demonstrating a quick, though not complete, programming application. Second, they can often be used by non-programmers. Applications generators are often used in programming voice processors. Applications generators come in many flavors. They may be general purpose tools. Or they may provide support for specific application segments such as: connecting voice response units to mainframe databases; voice messaging system development; audiotex system development, etc.

One of applications generators' biggest advantages is that their ability to translate user specified screens and menus into programming code. In essence you produce the screen or menu using an interface as simple as a word processor. Then the applications generator translates that screen into programming code in a language, such as the widely-used ""C." Once translated into C, a programmer proficient in C could go through the code and ""improve" on it.

APPLICATIONS PROCESSOR
A special purpose computer which attaches to a telephone system and allows the telephone system (and the people using it) to do different "applications" things, like voice mail, interactive voice response, etc. We think AT&T invented the term Applications Processor.

APPSERVER (AppServer)
A telephony server software program from Dialogic that enables an open SCSA telecomputing LAN server. Multiple applications from multiple vendors may control this server locally or remotely.

APPS GENS
See APPLICATIONS GENERATORS.

AT WORK PROTOCOL
Microsoft's integration protocol for linking peripheral office equipment, including telephony services, to Windows applications.

AUDIO
Sound you hear which may be converted to electrical signals for transmission. A human being who hasn't had his ears blown by listening to a Sony Walkman or a ghetto blaster can hear sounds from about 15 to 20,000 hertz.

AUDIO FREQUENCIES
Those frequencies which the human ear can detect (usually in the range of 15 to 20,000 hertz). Only those from 300 to 3,000 hertz are transmitted through the phone. Which is why the phone doesn't sound "Hi-Fi."

AUDIO MENU
Options spoken by a voice processing system. The user can choose what he wants done by simply choosing a menu option -- hitting a touchtone on his phone, or speaking a word or two. There are basically two ways of organizing computer or voice processing software -- menu-driven and non-menu driven. Menu-driven programs are easier for users to use. But they can only present as many options as can be reasonably spoken in a few seconds. Audio menus are typically played to callers in automated attendant/voice messaging, voice response and transaction processing applications. See also MENU.

AUDIO MESSAGING INTERCHANGE SPECIFICATION

AMIS. Issued in February 1990, AMIS is a series of standards aimed at addressing the problem of how voice messaging systems produced by different vendors can network or inter-network. Before AMIS, systems from different vendors could not exchange voice messages. AMIS deals only with the interaction between two systems for the purpose of exchange voice messages. It does not describe the user interface to a voice messaging system, specify how to implement AMIS in a particular systems or limit the features a vendor may implement.

AMIS is really two specifications. One, called AMIS-Digital, is based on completely digital interaction between two voice messaging systems. All the control information and the voice message itself, is conveyed between systems in digital form. By contrast, the AMIS-Analog specification calls for the use of DTMF tones to convey control information and transmission of the message itself is in analog form. AMIS was discussed in detail in the October 1990 issue of Business Communications Review. AMIS specifications are available from Hartfield Associates, Boulder CO. 303-442-5395.

AUDIO RESPONSE UNIT

A device which translates computer output into spoken voice. Let's say you dialed up a computer and it said "If you want the weather in Chicago, pushbutton "123," then it would give you the weather. But that weather would be "spoken" by an audio response unit.

AUDIOTEX

A generic term for voice response equipment and services. Audiotex may be passive or interactive. In "passive," the classic applications fall into "Jokes, Scopes and Soaps." Dial a number, hear the joke of the day. Dial a number, hear your horoscope. Dial a number, hear what's happening on your latest TV soap.

In its interactive form, Audiotex is basically another term for Interactive Voice Response. You call a phone number. A machine answers. It presents you several options, "Push 1 for information on Plays, Push 2 for information on movies, Push 3 for information on Museums." If you push 2, the machine may come back, "Push 1 for movies on the south side of town, Push 2 for movies on the north side of town, etc." Or it says, "To hear your present bank balance, touchtone your bank account number in now." See also **INTERACTIVE VOICE RESPONSE.**

AUDIOTEXT
An alternate, though wrong, saying of AUDIOTEX.

AUTO FAX TONE
Also called CNG, or Calling Tone. This tone is the sound produced by virtually all Group 3 fax machines when they dial another fax machine. CNG is a medium pitch tone (1100 Hz) that lasts 1/2 second and repeats every 3 1/2 seconds. A FAX machine will produce CNG for about 45 seconds after its dials. See also CNG.

AUTOMATED ATTENDANT
A specialized form of an Interactive Voice Response system. An IVR connected to a PBX. When a call comes in, this device answers it, and says something like: "Thanks for calling the ABC Company. If you know the extension number you'd like, pushbutton that extension now and you'll be transferred. If you don't know it, pushbutton "0" (zero) and the live operator will come on. Or wait a few seconds and the operator will come on, anyway." Sometimes the automated attendant might give you other options -- like "dial 3" for a directory. Automated attendants are sometimes connected also to voice mail systems ("I'm not here. Leave a message for me."). Some people react well to automated attendants. Others don't.

A good rule: Before you spring an automated attendant on your people/customers/subscribers, etc., let them know. Train them a little. They'll then perceive the automated attendant device a lot more positively. Tip: When you reach an automated attendant and don't know the extension number, dial the person's last name on your touchtone pad. Better automated attendants will recognize the name and translate it into the extension number.

AUTOMATIC CALL DISTRIBUTOR

ACD. A specialized phone system used for handling many incoming calls. Typically used by airlines, rent-a-car companies, hotels, etc. An ACD has four functions. First, it will recognize and answer an incoming call. Second, it will look in its database for instructions on what to do with that call. Third, based on that database's instructions, it will send the call to a recording that "somebody will be with you soon, please don't hang up!" Fourth, it will send the call to one operator of a group of operators -- as soon as that operator has completed their previous call, and/or the caller has heard the canned message. The term Automatic Call Distributor comes from distributing the incoming calls in some logical pattern to a group of operators. That pattern might be Uniform (to distribute the work uniformly). It also might be Top-down. Distributing calls logically is the function most people associate with an ACD. But it's not the most important. Much more important is the management information which the ACD produces. This information typically is of two sorts -- 1: The arrival of incoming calls (when, how many, which lines, from where, etc.) and 2: How many callers were put on hold, told to wait and didn't. This is called information on ABANDONED CALLS. This information is very important for staffing, for buying lines from the phone company and also for figuring what level of service to provide to the customer. And what different levels of service (how long for people to answer the phone) might cost.

Automatic Call Distributors are now combined with Interactive Voice Response Systems and ANI (automatic number identification) to give much effective and faster handling of incoming customer calls.

AUTOMATIC CALL SEQUENCER

ACS. A device for handling incoming calls. Typically it performs three functions. 1. It answers an incoming call, gives the caller a message, and puts them on "Hold." 2. It signals the agent (the person who will answer the call) which call on which line to answer. Typically, the call which it signals to be answered is the call which has been on "hold" the longest. 3. It provides management information, such as how many abandoned calls there were, how long the longest person was kept on hold, how long the average "on hold" was, etc.

AUTOMATIC CALLBACK

When a caller dials another internal extension and finds it busy, the caller dials some digits on his phone or presses a special "automatic callback" button. When the person he's calling hangs up, the phone system rings his number and the number of the original caller and the phone system automatically connects the two together. This feature saves a lot of time by automatically retrying the call until the extension is free.

AUTOMATIC CALLING UNIT

ACU. A device that places a telephone call on behalf of a computer.

AUTOMATIC CIRCUIT ASSURANCE

ACA is a PBX feature that helps you find bad trunks. The PBX keeps records of calls of very short and very long duration. If these calls exceed a certain parameter, the attendant is notified. The logic is that a lot of very short calls or one very long call may suggest that a trunk is hung, broken or out of order. The attendant can then physically dial into that trunk and check it.

AUTOMATIC NUMBER IDENTIFICATION

ANI. Equipment at your central office which recognizes the telephone number of the person making the call so that information about the call can be sent to the call accounting (i.e. billing) system. There has been some discussion about using ANI capability and sending calling information all over the country along with the call itself. This way the recipient will know who's calling before he/she picks up the phone and answers the calls. So far, this service apparently awaits more universal introduction of ISDN, more corporate demand, more standardization, more seminars or whatever. See ANI.

BACKPLANE

The high-speed communications line which individual components of a modern electronic system are connected to. For example, all the extensions of a PBX are connected to line cards (circuit boards), which slide into the PBX's cage. At the rear of the PBX cage, there is a connector. Each of these connectors is plugged into the PBX's Backplane. Also called a Backplane Bus. This backplane bus is typically very high speed, since it carries many conversations, much address information and considerable signaling. These days, the backplane bus is typically a time division multiplexed line -- somewhat like a train with many cars, each of which represent a time slice of another conversation. The backplane's capacity determines the overall capacity of the switch.

BASEBOARD

A Dialogic voice processing board without any daughterboards attached.

BATTERY

1. Term used to reference the DC power source of a telephone system. Often called "Talk Battery." 2. Storage battery used with central office switching systems and PBXs serving locations which cannot tolerate outages. Batteries serve the following purposes: Act as a filter across the generator or power rectifier output to smooth out the current and reduce noise; provide a cushion against periodic overloads exceeding the generator/rectifier capacity; supply emergency power for a limited time in event of commercial power failure.

BOARD

An SCSA defniition. Any hardware module that controls its own physical interface to the SCbus or SCxbus. From a programming point of view, a board is an addressable system component that contains resources.

BONG

A tone that long distance carriers and value added carriers make in order to signal you that they now require additional action on your part -- usually dialing more digits.

BUSY

In use. "Off-hook". There are slow busies and fast busies. Slow busies are when the phone at the other end is busy or off-hook. They happen 60 times a minute. Fast busies (120 times a minute) occur when the network is congested with too many calls. Your distant party may or may not be busy, but you'll never know because you never got that far.

C

C
The most common programming language in the voice processing industry. C operates under MS-DOS or UNIX, and other operating systems. It is very powerful and is becoming a standard also for programming telecom switches, PBX and central office.

CADENCE
In voice processing, cadence is used to refer to the pattern of tones and silence intervals generated by a given audio signal. Examples are busy and ringing tones. A typical cadence pattern is the US ringing tone, which is one second of tone followed by three seconds of silence. Some other countries, such as the UK, use a double ring, which as two short tones within about a second, followed by a little over two seconds of silence.

CALL CENTER
A place where calls are answered and calls are made. A call center will typically have lots of people (also called agents), an automatic call distributor, a computer for order-entry and lookup on customers' orders. A Call Center could also have a predictive dialer for making lots of calls quickly. The term "call center" is broadening. It now includes help desks and service lines.

CALL CONTROL
The establishment and control of telephone links. This involves answering, originating, transferring, conferencing, and terminating telephone calls, and monitoring their status. The electronic signaling functions that control the setting up, monitoring and tearing down of telephone calls.

CALL PROCESSING
Call processing is setting up, connecting, transferring, disconnecting, etc. a phone call. Call processing does not affect the content -- voice or otherwise -- of the conversation. That process is called "voice processing." And that is a broader concept, encompassing everything from compression, storage, editing to recognition.

CALL PROGRESS TONE
A tone sent from the telephone switch to tell the caller of the progress of the call. Examples of the common ones are dial tone, busy tone, ringback tone, error tone, re-order, etc. Some phone systems provide additional tones, such as confirmation, splash tone, or a reminder tone to indicate that a feature is in use, such as confirmation, hold reminder, hold, intercept tones.

CALLER ID (caller ID)
Signals that identify the telephone number of the calling party; these signals are sent along the telephone line at the start of the call. Not all telephone network connections support caller ID signaling. In a more general sense, caller ID can be obtained by prompting the caller for an identity code.

CALLING PARTY IDENTIFICATION
A new service being tested in some local areas which tells the person being called which number is calling them. They can then decide to answer or not answer the call. See ANI, which stands for Automatic Number Identification.

CALLPATH
IBM's announced telephone system link to IBM's computers. See CALLPATH SERVICES ARCHITECTURE.

CALLPATH CICS
Enabling software that connects your telephone systems with your IBM 370 or 390 (i.e. the mainframe version of CallPath/400, which works on the AS/400 platform).

CALLPATH HOST
IBM and ROLM's CICS-based integrated voice and data applications platform which links to ROLM's 9751 PBX. See CALLPATH SERVICES ARCHITECTURE.

CALLPATH SERVICES ARCHITECTURE
CSA. IBM's program for intergrating voice and data technologies. It is both a strategy and set of open architecture commands and interfaces for integrating voice and database technologies. The idea is that with CallPath a call will arrive at a computer terminal simultaneously with the database record of the caller. And such call and database record can be transferred simultaneously to an expert, a supervisor, etc. The first implementation is CallPath/400 which works with the IBM AS/400 minicomputer. CallPath CICS works with IBM mainframes. See OPEN APPLICATION INTERFACE and DIRECTTALK.

CAS
Communicating Applications Specification. A high-level API (application programming interface) developed by Intel and DCA that was introduced in 1988. CAS enables software developers to integrate fax capability and other communication functions into their applications.

CENTREX
Centrex is a business telephone service offered by a local telephone company from a local central office. Centrex is basically single line telephone service delivered to individual desks (the same as you get at your house) with features, i.e. "bells and whistles," added. Those "bells and whistles" include intercom, call forwarding, call transfer, toll restrict, least cost routing and call hold (on single line phones). Centrex is known

by many names among operating phone companies, including Centron and Cenpac. Centrex comes in two variations -- CO and CU. CO means the Centrex service is provided by the Central Office. CU means the central office is on the customer's premises.

CHANNEL

1. An SCSA definition. A transmission path on the SCbus or SCxbus Data Bus that transmits data between two end points. 2. Typically what you rent from the telephone company. A voice-grade transmission facility with defined frequency response, gain and bandwidth. Also, a path of communication, either electrical or electromagnetic, between two or more points. Also called a circuit, facility, line, link or path.

CHANNEL BANK

A multiplexer. A device which puts many slow speed voice or data conversations onto one high-speed link and controls the flow of those "conversations." Typically the device that sits between a digital circuit -- say a T-1 -- and a couple of dozen voice grade lines coming out of a PBX. One side of the channel bank will be connections for terminating two pairs of wires or a coaxial cable -- those bringing the T-1 carrier in. On the other side are connections for terminating multiple tip and ring single line analog phone lines or several digital data streams. Sometimes you need channel banks. Sometimes, you don't. For example, if you're shipping a bundle of voice conversations from one digital PBX to another across town in a T-1 format -- and both PBXs recognize the signal -- then you will probably not need a channel bank. You'll need a Channel Service Unit (CSU). If one, or both, of the PBXs is analog, then you will need a channel bank at the end of the transmission path whose PBX won't take a digital signal. See CHANNEL SERVICE UNIT and T-1.

CHANNEL SERVICE UNIT

CSU. A device used to connect a digital phone line (T-1 or less) coming in from the phone company to either a multiplexer, channel bank or directly to another device producing a digital signal, e.g. a digital PBX, or data communications device. A CSU performs certain line-conditioning, and equalization functions, and responds to loopback commands sent from the central office. A channel service unit is also called a Data Service Unit. See CSU.

CLASS

1. Custom Local Area Signaling Services. It is based on the availability of channel interoffice signaling. Class consists of number-translation services, such as call-forwarding and caller identification, available within a local exchange of Local Access and Transport Area (LATA). CLASS is a service mark of Bellcore. Some of the phone services which Bellcore promotes for CLASS are Automatic Callback, Automatic Recall, Calling Number Delivery, Customer Originated Trace, Distinctive Ringing/Call Waiting, Selective Call Forwarding and Selective Call Rejection. See also CALLING LINE IDENTIFICATION.
2. In an object-oriented programming environment, a class defines the data content of a specific type of object, the code that manipulates it, and the public and private programming interfaces to that code.

CLASS 1

The Class 1 interface is an extension of the EIA/TIA's (Electronics Industry Association and the Telecommunications Industry Association) specification for fax communication, known as Group III. Class I is a series of Hayes AT commands that can be used by software to control fax boards. In Class 1, both the T.30 (the data packet creation and decision making necessary for call setup) and ECM/BFT (error-correction mode/binary file transfer) are done by the computer.

A specification being developed (fall of 1991) Class 2, will allow the modem to handle these functions in hardware. Industry analysts believe Class 2 will be the standard for the long haul, but approval is slow. Even so, some modem makers will shortly deliver data/fax modems. See also CLASS 1 OFFICE.

CLEAR CHANNEL
A channel which is used exclusively for data transmission, with no bandwidth required for administrative messages such as signaling or synchronization. All SCbus data channels are clear.

CNG
Also called Auto Fax Tone, or Calling Tone. This tone is the sound produced by virtually all fax machines when they dial another fax machine. CNG is a medium pitch tone (1100 Hz) that lasts 1/2 second and repeats every 3 1/2 seconds. A fax machine will produce CNG for about 45 seconds after it dials. The CNG tone is useful for owners of fax/phone/modem switches. Such switches answer an incoming call. If they hear a CNG tone, they will transfer the call to a fax machine. If they don't, they'll transfer the call to a phone, answering machine or perhaps a modem. Depends on how they're set up.

CO
Central Office. In North America, a CO is that location which houses a switch to serve local telephone subscribers. Sometimes the words "central office" are confused with the switch itself. In Europe and abroad, the words "central office" are not known. The more common words are "public exchange." But those words tend to refer more to the switch itself, rather than the site, as in North America. See also CENTRAL OFFICE or PUBLIC EXCHANGE.

CO LINES
These are the lines connecting your office to your local telephone company"s Central Office which in turn connects you to the nationwide telephone system.

CO SIMULATOR
A desktop device which pretends to act like a mini-central office. The smallest version will consist of two lines and two RJ-11 jacks. Plug a phone into both jacks. Pick up one phone. You hear dial tone. Dial or touchtone two or three digits. Bingo, the second phone rings. You pick up the second phone. You can have a conversation with yourself or with a machine -- like a voice processing system. Most central office simulators can simulate normal on-hook, off-hook, dialing, answering, speaking, etc. Some now can simulate caller ID features -- including number of person calling.

COMPILER
A computer program used to convert symbols meaningful to a human into codes meaningful to a computer. For example, taking instructions written in a "higher" level language such as C, BASIC, COBOL or ALGOL and converting them into machine code which can be read and acted upon by a computer.

COMPRESSION ALGORITHM
The arithmetic formulae which convert a signal into smaller bandwidth or few bits.

COMPUTER-AIDED DIALING
Another term for Predictive Dialing. See PREDICTIVE DIALING.

CONFERENCE BRIDGE
A telecommunications facility or service which permits callers from several diverse locations to be connected together for a conference call. The conference bridge contains electronics for amplifying and balancing the conference call so everyone can hear each other and speak to each other. The conference call's progress is monitored through the bridge in order to produce a high quality voice conference and to maintain decent quality as people enter or leave the conference.

CPE
Customer Provided Equipment, or Customer Premise Equipment. Originally it referred to equipment on the customer's premises which had been bought from a vendor who was not the local phone company. Now it simply refers to telephone equipment -- key systems, PBXs, answering machines, etc. -- which reside on the customer's premises. "Premises" might be anything from an office to a factory to home.

CPU
The Central Processing Unit. The computing part of a computer. The "brain" of the computer. It manipulates data and processes instructions coming from software or a human operator. See CENTRAL PROCESSING UNIT.

CSU
Channel Service Unit. Also called a Data Service Unit. A device to terminate a digital channel on a customer's premises. It performs certain line-conditioning, and equalization functions, and responds to loopback commands sent from the central office. A CSU sits between the digital line coming in from the central office and devices such as channel banks or data communications devices. See CHANNEL SERVICE UNIT.

CTI
Computer Telephone Integration. Basically a polite term for connecting a computer to a telephone switch and having the computer issue the switch commands to move calls around. The most classic application for CTI is in call centers. Picture this: A call comes in. That call carries some form of caller ID -- either ANI or Class Caller ID. The switch "hears" the calling number, strips it off, sends it to the computer. The computer does a lookup, sends back the switch instructions on what to do with the call. The switch follows orders. It might send the call to a specialized agent or maybe just to the agent the caller dealt with last time.

CUSTOMER PREMISES EQUIPMENT
CPE. Terminal equipment, supplied by either the telephone common carrier or by a competitive supplier, which is connected to the nationwide telephonenetwork and resides on the customer's premises.

CVSD
Continuously Variable Slope Delta modulation. A method for coding analog voice signals into digital signals.

D/A
Digital to analog conversion.

dB
Decibel. A unit of measure of signal strength, usually the relation between a transmitted signal and a standard signal source. Therefore, 6 Db of loss would mean that there is 6 Db difference between what arrives down a communications circuit, and what was transmitted by a standard signal generator. See DECIBEL for a better explanation.

DDS
Dialogic's Device Driver Server.

DEVICE DRIVER
A piece of software that expands MS-DOS's ability to work with peripherals. The routines that make peripherals (a mouse, a RAM disk, a print spooler) work. They may be part of another program (many applications include device drivers for printers) or separate programs. DOS comes with two device drivers: ANSI.SYS and VDISK.SYS. These are things DOS is not used to. The first is to expand the capabilities of the keyboard. The second is used to carve a RAM disk out of the 640K of RAM that MS-DOS can address. In short, device drivers can be used to expand the abilities of DOS. They are loaded from your CONFIG.SYS file when you boot up the computer. Some network operating systems run as device drivers. They do this to get loaded before anything else so they can get the memory space they want. By being loaded as a device driver they are able to tell DOS how to operate certain "network" things, like virtual drives, shared printers, spoolers, print queues and so on. See DRIVER and VOICE DRIVER.

DIAL-IT 900 SERVICE
A special one-way mass calling service that allows prospects, customers and others to reach you from anywhere across the country. In contrast to 800-service, the caller pays the 900 charge. The caller generally pays one charge for the first minute, with a lesser charge for each additional minute. DIAL-IT 900 Service is a great way to involve your customers and prospects in a promotion! Premium Billing lets you select a rate above standard DIAL-IT 900 rates. The long distance carriers (through their deals with local phone companies) handle the billing. You, the information provider, may split the revenues with the long distance provider. International DIAL-IT 900 service is currently available from a growing number of countries.

DIAL PULSING

A means of signaling consisting of regular momentary interruptions of a direct or alternating current at the sending end in which the number of interruptions corresponds to the value of the digit or character. In short, the old style of rotary dialing. Dial the number "five" and you'll hear five "clicks." See DIAL SPEED and DTMF.

DIAL SPEED

The number of pulses that a rotary dial can send in a given period of time, typically 10 pulses per second. A Hayes modem with a communications package, like Crosstalk, can send 20 pulses a second.

DIAL TONE

The sound you hear when you pick up a telephone. Dial tone is a signal (350 + 440 Hz) from your local telephone company to you that it is alive and ready to receive the number you dial. If you have a PBX, dial tone will typically be provided by the PBX. Dial tone does not come from God or the telephone instrument on your desk. It comes from the switch to which your phone is connected to.

DICTATION ACCESS AND CONTROL

A telephone system feature which allows a user to dial a dictation machine and use that machine (giving it instructions by pushbutton) as if that machine were in his office. Typically the material on that dictation machine is taken off by one or several typists out of a centralized word processing pool and word processed into letters, reports, legal briefs, etc. Telephone suppliers usually don't supply the dictation equipment. Newer telephone dictation machinery is, in reality, a specialized application of voice processing equipment. See VOICE PROCESSING.

DID
Direct Inward Dialing. You can dial inside a company directly without going through the attendant. This feature used to be an exclusive feature of Centrex but it can now be provided by virtually all modern PBXs and some modern hybrids.

DIGITAL RECORDING
A system of recording in which musical information is converted into a series of pulses that are translated into a binary code intelligible to computer circuits, and stored on magnetic tape of magnetic discs. Also called PCM - Pulse Code Modulation.

DIGITAL TRUNK
Generic name for a telephone connection for which uses digital rather than analog transmission. Common examples are T-1 in North America and E-1 in Europe.

DIGITAL-TO-ANALOG CONVERSION
A circuit that accepts digital signals and converts them into analog signals. A modem typically has such a circuit. It also has other circuits, such as those doing with signaling. See MODEM.

DIGITIZE
Converting an analog or continuous signal into a series of oncs and zeros, i.e. into a digital format.

DIGITIZED VOICE
Analog voice signals represented in digital form. There are many ways of digitizing voice. See Pulse Code Modulation for the most common.

DIRECT INWARD DIALING
DID. The ability for a caller outside a company to call an internal extension without having to pass through an operator or attendant. In large PBX systems, the dialed digits are passed down the line from the

CO (central office). The PBX then completes the call. Direct Inward Dialing is often proposed as Centrex's major feature. But automated attendants (a specialized form of interactive voice response systems) also provide a similar service.

DIRECT TALK (DirectTalk)

A family of IBM voice processing products introduced in the summer of 1991. According to IBM, its "new IBM CallPath DirectTalk product line lets businesses automate routine operations and also provide callers with easy access to many kinds of information over the telephone -- at any hour of the day and with greater accuracy. Businesses can raise the level of service they provide and do it with fewer people and with greater efficiency." According to IBM's press release, the North Carolina Employment Security Commission is using the DirectTalk product to automate the filing of unemployment benefit claims. For several years, the state has allowed claimants to verify their unemployment status by answering a series of questions on a postcard. Based on initial results, Claimants who use the telephone to answer the questions can receive checks up to four days sooner than if they answer by mail.

DISCONNECT SUPERVISION

The change in electrical state from off-hook to on-hook. This indicates that the transmission connection is no longer needed.

DMX

Digital Matrix Switch. Dialogic board that provides digital switching among four PEB spans -- for a maximum of 96 channels.

DNIS

Dialed Number Identification Service. DNIS is a feature of 800 lines. Let's say you subscribe to several 800 numbers. You use one line for testing your advertisements on TV stations in Phoenix; another line for testing your advertisements on TV stations in Chicago; and yet another for Milwaukee. Now you get an automatic call distributor and you terminate all the lines in one group on your ACD. You do that because it's cheaper to man and run one group of incoming lines. One queue is more efficient than several small ones, etc. You have all your people answering all the calls. You now need to know which calls are coming from where. So your long distance carrier sends you the call's DNIS -- the numbers the person dialed to reach you. Those DNIS digits might come to you in many ways, depending on the technical arrangement you have with your long distance company. In-band or out-of-band. ISDN or data channel, etc. Make sure you understand the difference between DNIS and ANI. DNIS tells you the number your caller called. ANI is the number your caller called from.

DOS
Disk Operating System. As in MS-DOS, which stands for MicroSoft Disk Operating System. DOS is the software that organizes how a computer reads, writes and reacts with its disks -- floppy or hard -- and talks to its various input/output devices, including keyboards, screens, serial and parallel ports, printers, modems, etc.

DRIVER
Also called a Device Driver. A piece of software which lets applications software tell a computer's various parts what to do. Those parts might be everything from Dialogic boards to laser printers.

DROP AND INSERT
That process wherein a part of the information carried in a transmission system is demodulated (dropped) at an intermediate point and different information is entered (inserted) for subsequent transmission.

DRY T-1
T-1 with an unpowered interface.

DS-0
Digital Service, level 0. It is 64,000 bps, the worldwide standard speed for digitizing one voice conversation. There are 24 DS- channels in a DS-1.

DS-1
Digital Service, level 1. It is 1.544 megabits per second in North America, 2.048 Mbps elsewhere. Why there's no consistency is one of those wonderful questions. The 1.544 standard is an old Bell System standard. The 2.048 standard is a CCITT standard. Standard for 1.544 per second is 24 voice conversations. Standard for 2.048 is 30 conversations.

DSP
Display System Protocol or Digital Signal Processor. A Digital Signal Processor is a specialized computer chip designed to perform speedy and complex operations on digitized waveforms. Useful in processing sound and video.

DTC
Digital Trunk Controller.

DTMF
Dual Tone Multi Frequency. A fancy term for describing push button or Touchtone dialing. (Touchtone is a registered trademark of AT&T.) In DTMF, when you touch a button on a pushbutton dial, it makes a tone, which is actually the combination of two tones, one high frequency and one low frequency. Thus the name Dual Tone Multi Frequency. In U.S. telephony, there are actually two types of DTMF signaling -- one that is used on normal business or home pushbutton/touchtone phones, and one that is used for signaling within the telephone network itself. When you go into a central office, look for the testboard. There you'll see what looks like a standard touchtone pad. Next to the pad there'll be a small toggle switch that allows you to choose the sounds the touchtone pad will make -- either normal touchtone dialing or the network version.

The eight possible tones that comprise the DTMF signaling system were specially selected to easily pass through the telephone network without attenuation and with minimum interaction with each other. Since these tones fall within the frequency range of the human voice, additional considerations were added to prevent the human voice from inadvertantly imitating or "falsing" DTMF signaling digits. One way this was done to break the tones into two groups, a high frequency group and a low frequency group. A valid DTMF tone has only one tone in each group. Here is a table of the DTMF digits with their respective frequencies. One Hertz (abbreviated Hz.) is one cycle per second of frequency.

Digit	Low frequency	High frequency
1	697 Hz.	1209 Hz.
2	697	1336
3	697	1477
4	770	1209
5	770	1336
6	770	1477
7	852	1209
8	852	1336
9	852	1477
0	941	1336
*	941	1209
#	941	1477

There are four other digits defined in the DTMF system and usable for specialized applications that cannot be generated by standard telephones. They are:

Digit	Low frequency	High frequency
A	697 Hz	1633 Hz.
B	770	1633
C	852	1633
D	941	1633

DTMF CUT-THROUGH

The capability of a voice response system to receive DTMF tones while the voice synthesizer is delivering information, i.e. during speech playback. This capability of DTMF cut-through saves the user waiting until the machine has played the whole message (which typically is a menu with options). The user can simply touchtone his response anytime during the message -- when he first hears his selection number, when the message first starts, etc. When the voice processor hears the touchtoned selection (i.e. the DTMF cut-through), it stops speaking and jumps to the chosen selection. For example, the machine starts to say, "If you know the person you're calling, touchtone his extension in now.." But before you hear the "If you know" you pushbutton in 230, which you know is Joe's extension. Bingo, the message stops and Joe's extension starts ringing.

DUMB SWITCH

A slang word for a telecommunications switch that contains only basic switching software and relies on instructions sent it by an outside computer. Those instructions are typically fed the "dumb" switch through a cable from the computer to one or more RS-232 serial ports which the dumb switch sports. The switch makes no demands on what type of computer it talks to, but simply insists that it be able to feed the computer questions and promptly receive responses in a form that it (the switch) can understand. Plain ASCII is OK. For example, the dumb switch might signal the computer, "A call is coming in on port 23, what do I do now?" The computer might reply "Answer it and transfer it to extension 23." Or it might say "answer it and put it on hold," or "answer it, put it on hold and play it recording number three." In essence, a dumb switch is anything but. It is in reality an empty cage containing whatever network interface cards the user has chosen. Each of these network interface cards is designed to "talk" to one type of telephone line. That line might be a T-1 line. It might be a normal tip and ring loop start line. It might be a tie trunk with E&M signaling. The card may handle one or many lines, but always of the same type. The card knows how to answer a call or pulse out a call on that particular type of line. It has all the telephony smarts. What it lacks is the intelligence of what to do with the calls. That is provided by the outside computer. Well, almost. Most "dumb" switches do contain rudimentary intelligence -- a small computer and some memory. That computer is usually programmed to handle "default" calls -- and to handle calls should the link to the outside computer fail, or the outside computer itself fail. Dumb switches come in flavors all the way from residing in their own cabinet to being printed circuit cards which reside in one or more of the personal computer's slots. Dumb switches are programmed to do "specialized" telecom applications, for example emergency 911, added value 800 services, cellular switching, automatic call distributors, predictive dialers, etc. They can, of course, be programmed to be "normal" PBXs. The question increasingly being asked is "If I want to program a specialized telecom application should I use a dumb switch or should I use an open PBX?" And the answer is "It depends." Depends on what you want to do. Depends on what software is available, etc. See also OAI.

800 SERVICE
See 800 SERVICE, spelled as EIGHT HUNDRED SERVICE.

E & M LEADS
The pair of wires carrying signals between trunk equipment and a separate signaling equipment unit. The "M" lead transmits a ground or battery conditions to the signaling equipment. The "E" lead receives open or ground signals from the signaling equipment. These leads are also known as Ear and Mouth Leads. The Ear lead typically means to receive and the Mouth lead typically means to transmit. Changes of voltage on these leads convey such information as seizure of circuit, recognition of seizure, release of circuit, dialed digits, etc. In the old days it was the

PBX operators who originated trunk calls by asking the long distance carrier for free trunks using their mouth or M lead. If the carrier had a free trunk, the PBX heard about it through its ear or E lead. See also E & M SIGNALING.

E & M SIGNALING
In telephony, an arrangement that uses separate leads, called respectively the "E" lead and "M" lead, for signaling and supervisory purposes. The near end signals the far end by applying -48 volts dc (vdc) to the "M" lead, which results in a ground being applied to the far end's "E" lead. When -48 vdc is applied to the far end "M" lead, the near-end "E" lead is grounded. The "E" originally stood for "ear," i.e., when the near-end "E" lead was grounded, the far end was calling and "wanted your ear." The "M" originally stood for "mouth," because when the near-end wanted to call (i.e., speak to) the far end, -48 vdc was applied to that lead.

When a PBX wishes to connect to another PBX directly or to a remote PBX or extension telephone over a leased voice grade line, a channel on T-1, the PBX uses a special line interface which is quite different from that which it uses to interface to the phones it's attached directly to (i.e. with in-building wires). The basic reason for the difference between a normal extension interface and the long distance interface is that the signaling requirements differ -- even if the voice signal parameters such as level and two-wire, 4-wire remain the same. When dealing with tie lines or trunks it is costly, inefficient and too slow for a PBX to do what an extension telephone would do, i.e. go off hook, wait for dial tone, dial, wait for ringing to stop, etc. The E&M tie trunk interface device is the closest thing there is to a standard that exists in the PBX, T-1 multiplexer, voice digitizer telco world. But even then it comes in at least five different flavors. See E & M LEADS.

EIGHT HUNDRED SERVICE
800-Service. A generic and common (and not trademarked) term for AT&T's, MCI's, US Sprint's and the Bell operating companies' IN-WATS service. All these IN-WATS services have "800" as their "area code."

Dialing an 800-number is free to the person making the call. The call is billed to the person or company being called. The telephone company suppliers of 800 services use various ways to configure and bill their 800-services. One way: you can buy an 800 line which will ring on your normal phone line. You'll only pay per call, but you won't receive any incoming call if you're making an outgoing one. (You can even terminate an 800 number on your cellular phone.) For other 800 services you might pay a flat monthly rate plus "so-much" (i.e. timed usage) per call. That timed usage may include some calculation for the distance the incoming call traveled. 800-Service is now available for calls from Canada and some countries in Europe. More and more long distance companies are introducing 800-service.

800 Service works like this: You're somewhere in North America. You dial 1-800 and seven digits. Your local central office sees the "1" and recognizes the call as long distance. It ships that call to a bigger central office (or perhaps processes the call itself). At that central office it's processed, a machine will recognize the 800 "area code" and examine the next three digits. Those three digits will tell which long distance carrier to ship the call to. Each long distance company has been assigned specific 800 three digit "exchanges." For example, MCI has the exchange 999. AT&T has the exchange 542. If you want a phone number beginning with 800-999, then you must subscribe to MCI 800 service. If you want a phone number beginning with 800-542, you must subscribe to AT&T 800 service. Once the 800-call is "passed off" to whichever carrier it belongs to, that carrier sends the call to a switch attached to a huge "translation" database. The call arrives at the switch. The database says the call 800-NNN-XXXX is really 212-555-1234, and sends the 800 line to that number.

As a real-life example, Telecom Library, publishers of this book, has an 800 number, namely 800-LIBRARY (or 800-542-7279). When you call that number, the following number -- 212-206-6870 -- in New York City rings. Dialing 800-542-7279 is effectively the same as dialing 212-206-6870, except that if you dial 212-206-6870 you'll pay. If you dial 800-

542-7279, I'll pay. Because 800 long distance service is essentially a database lookup and translation telephone service, there are endless "800 services" you can create.

E-1
Another name given to the CEPT digital telephony format devised by the CCITT that carries data at the rate of 2.048 Mbps (DS-1 level). CEPT format consists of 30 voice channels, one signaling channel, and one framing (synchronization) channel. Since robbed-bit signaling is not used (as it is for T-1 in North America) all 8 bits per channel are used to code the waveshape sample. E-1 is the European version of North American T-1, though T-1 is 1.544 Mbps. See T-1.

ELECTRONIC VOICE MAIL
A system which stores messages spoken by a user usually over a telephone, which can be retrieved by the intended recipient when that person next calls into the system. Also called Voice Mail, it operates just like a touch-tone controlled answering machine.

EMAIL-TO-FAX (Email-to-fax)
Conversion of electronic mail (ASCII text) into fax image format, suitable for sending to a fax machine.

EMULATE
To imitate a computer or computer system by a combination of hardware and software that allows programs written for one computer or terminal to run on another. The most common data terminal is a DEC VT-100. Our communications program, Crosstalk, allows us to "emulate" a DEC-VT100 on our IBM PCs and PC clones.

ESP
Enhanced Service Provider. A vendor who adds value to telephone lines using his own provided software and hardware. Also called an IP, or Information Provider. An example of an ESP is a public voice mail box provider or a database provider, say one giving the latest airline fares. An

ESP is an American term, unknown in Europe, where they're most called VANs, or Value Added Networks. See also INFORMATION PROVIDER.

EVENT
An unsolicited, asynchronous signal from a device or device driver that reports on a change in the status of the device. Events are generally attention-getting messages, allowing a process to know when a task is complete or when an external event occurs.

EXPANSION SLOTS
In a computer there are card slots for adding accessories such as voice processing interface cards, internal modems, extra drivers, hard disks, monitor adapters, hard disk drivers, etc. Most modern PBXs are actually cabinets with nothing but expansion slots. And into these slots we fit trunk cards, line cards, console cards, etc. Some phone systems have "universal" slots, meaning you can put any card in any slot. Some phone systems have dedicated expansion slots, meaning that they expect only a certain card in that slot.

FAST BUSY
A busy signal which sounds at twice the normal rate (120 interruptions/minute vs. 60/minute). A "fast busy" signal indicates all trunks are busy.

FAX-BACK
You go to your fax machine. You dial a voice response unit. It says "Punch in 23 if you want to receive the latest specs on our new XYZ machine. Now when you are ready, hit the 'start' button on your fax machine and hang up." A few seconds later, your fax machine disgorges

paper containing the latest specs on the XYZ machine. Welcome to "fax-back."

FAX-MAILBOX
Companies can send facsimiles of documents to be stored for later retrieval to a fax mailbox -- a cousin to a voice mailbox. Travelers can check their fax mailboxes and have the faxes sent to convenient locations, like a hotel front desk.

FAX-SERVER
A specialized Interactive Voice Response system which sends facsimile messages to a fax machine you designate by touchtoning in numbers. Picture a database -- not of files, but of pages of facsimile text. You call TELECONNECT Magazine's phone number. You ask to receive an article published on Voice Processing in the January, 1989 issue. The interactive voice response system asks you for your fax machine's phone number. You hang up. Within seconds your fax machine is receiving a fax of the article you requested.

FCC REGISTRATION NUMBER
A number assigned to specific telephone equipment registered with the FCC, as set forth in FCC docket 19528, part 68. The presence of this number affixed to a device indicates that the FCC has approved it as being a compatible device for direct connection to telephone line facilities.

FIRMWARE
Programs kept in semipermanent storage, such as various types of read-only memory. These programs can be altered, but with difficulty. Firmware is used in conjunction with hardware and software. It also shares the characteristics of both. Firmware is usually stored on PROMS (Programmable Read Only Memory) or EPROMs (Electrical PROMS). Firmware contains software which is so constantly called upon by a computer or phone system that it is "burned" into a chip, thereby becoming Firmware. The computer program is written into the PROM

electrically at higher than usual voltage, causing the bits to "retain" the pattern as it is "burned in". Firmware is non-volatile. It will not be "forgotten" when the power is shut off. Hand-held calculators contain firmware with instructions for doing their various mathematical operations.

FLASH

Quickly depressing and releasing the plunger in or the actual handset-cradle to create a signal to a PBX or Centrex that special instructions will follow such as transferring the call to another extension.

FLASH HOOK

A brief on-hook period. A common use of the flash-hook is in the residential call waiting feature. A new call comes in while a conversation is in progress. The central office, however, doesn't give the caller a busy signal. The caller hears ringing and the called party, who's having a conversation, hears a special "beep" tone. If he or she chooses, the called party can quickly push the hook-switch on their telephone, sending what the industry calls a "flash hook." The central office puts the original caller on hold and the new dial tone, allowing a three-way conference or call transfer.

FLOWCHART

A graphic or diagram which shows how a complex operation, e.g. programming, takes place. The flowchart breaks that operation down into its smallest, and easiest-to-understand events.

FOD

Fax On Demand. Dial up a number. hear an voice response unit say, "Would you like a timetable?" You say "Yes." It says put in your fax number. You punch in your fax number and bingo, your fax machine starts receiving a fax of the timetable.

FRAME, DATA
A set of time slots which are grouped together for synchronization purposes. For example, as with SCSA, the number of time slots in each frame depends on the SCbus or SCxbus Data Bus data rate. Each frame has a fixed period of 125us. Frames are delineated by the timing signal FSYNC.

FRAME, MESSAGE
A data link layer frame the encapsulates control and signaling data transmitted on hte SCbus or SCxbus Message Bus. The form of a Message Bus frame is fully compliant with ISO HDLC UI (Unnumbered Information) Frame specifications.

FREQUENCY RESPONSE
The variation (dB) in relative strength between frequencies in a given frequency band, usually the voice frequency band of an analog telephone line.

GLARE
Glare occurs when both ends of a telephone line or trunk are seized at the same time for different purposes or by different users. Most embarrassing.

GROUND START
A way of signaling on subscriber trunks in which one side of the two wire trunk (typically the "Ring" conductor of the Tip and Ring) is momentarily grounded to get dialtone. There are two types of switched trunks one can typically lease from a local phone company -- ground start and loop start. PBXs work best on ground start trunks, though many will work -- albeit intermittently -- on both types. Normal single line phones

and key systems typically work on loop start lines. You must be careful to order the correct type of trunk from your local phone company and correctly install your telephone system at your end -- so that they both match. In technical language, a ground start trunk initiates an out-going trunk seizure by applying a maximum local resistance of 550 ohms to the tip conductor. See LOOP START.

HDLC
High-level Data Link Control: an ISO standard, bit oriented, data link layer protocol used for the transmission of data and messages on the SCBus and SCxbus.

HIGH LEVEL LANGUAGES
Essentially any of the computer languages whose code is not unique to the hardware or architecture of a particular computer. High level languages are more like human language than the machine language which computers talk. High level languages translate human instructions into the machine language computers can understand, but which humans don't have to (in order to tell the computer what to do). Computer languages such as C, BASIC, FORTRAN, COBOL and Pascal are high level languages. They are a number of levels (at a High Level) away from the actual bit manipulation (machine language, also called "bit twiddling" by the Hackers).

HIVR
Host Interactive Voice Response. Tieing a voice response unit into a mainframe computer which has lots of data. Applications which can be produced include bank-by-phone, reservations-by-phone, etc.

HOOK FLASH
See FLASH

HOOKSWITCH
Also called SWITCHHOOK. It's typically the place on your telephone instrument where you lay your handset. When you lift the handset, you are said to be going "off hook." When you place your handset back, you are said to be "on hook." When you lift the handset (i.e. go off hook) you are, in effect, signaling your central office that you'd like to make a call, or that you have answered the incoming ringing call.

HOST COMPUTER
A computer attached to a network providing primarily services such as computation, data base access or special programs of special programming languages. The central computer in a time-sharing operation.

HOST INTERACTIVE VOICE RESPONSE
Interactive Voice Response system communicating with a host computer (typically a mainframe computer). See HIVR.

HOST PROCESSOR
Same as HOST COMPUTER.

HUNT
Refers to the progress of a call reaching a group of lines. The call will try the first line of the group. If that line is busy, it will try the second line, then it will hunt to the third, etc. See also HUNT GROUP.

HUNT GROUP
A series of telephone lines organized in such a way that if the first line is busy the next line is hunted and so on until a free line is found. Often this arrangement is used on a group of incoming lines. Hunt groups may start with one trunk and hunt downwards. They may start randomly and hunt in clockwise circles. They may start randomly and hunt in counter-clockwise circles. Inter-Tel uses the terms "Linear, Distributed and Terminal" to refer to different types of hunt groups. In data

communications, a hunt group is a set of links which provides a common resource and which is assigned a single hunt group designation. A user requesting that designation may then be connected to any member of the hunt group. Hunt group members may also receive calls by station address.

HYPERCHANNEL
A data path on the SCbus or SCxbus Data Bus made up of more than one time slot. By bundling time slots into a hyperchannel, data paths with a bandwidth greater than 64 Kbps can be created.

IDLE
A state of the SCbus or SCxbus Message Bus where no information is being transmitted and the bus line is pulled high.

IN-BAND SIGNALING
Signaling made up of tones which pass within the voice frequency band, and are carried along the same circuit as the talk path that is being established by the signals. Virtually all signaling -- request for service, dialing, disconnect, etc. -- in the U.S. today is in-band signaling. Most of that signaling is MF -- multi-frequency dialing. The more modern form of signaling is out-of-band.

INFORMATION CENTER MAILBOXES
A voice bulletin board on a voice mail system. Here's their explanation: Multiple callers can access, directly or indirectly, recorded announcements containing information that would otherwise have been given live by employees. Callers are frequently "outside" users of the system. One type of "listen only" mailbox simply plays the messages to the callers. This technology, sometimes known as audiotex, makes it possible to create a verbal database so callers can select which *information*

they want to hear. Another type of Information Center Mailbox prompts callers to reply to announcements. Callers wanting further information can be given the opportunity to leave their names and phone numbers after listening to a product description. They can also be transferred to a designated employee who can immediately take an order. If desired, a password can be required before confidential or controlled access information can be heard.

INFORMATION PROVIDER
A business or person providing information to the public for money. The information is typically selected by the caller through touchtones, delivered using voice processing equipment, and transmitted over tariffed phone lines, e.g., 900, 976, 970. Typically, billing for information providers' services is done by a local or long distance phone company. Sometimes the revenues for the service are split by the information provider and the phone company. Sometimes the phone company simply bills a per minute or flat charge. A typical "information provider" is American Express, which provides a service -- 1-900-WEATHER. By dialing that number you can touchtone in city names and find out temperatures, weather forecasts, etc. Calling 1-900-WEATHER costs several dollars a minute.

INTERACTIVE VOICE RESPONSE
IVR. Picture a standard personal computer. You enter information through the keyboard. You see the results of your work on your screen. With interactive voice response, the keyboard, (or data entry device) becomes the touchtone pad of a telephone (either local, but most likely remote) and the screen (the output device) becomes a voice synthesizer which electronically converts the computer's output (what you'd normally see on the screen) into spoken words. You call a phone number. A machine answers. It presents you with several options, "Push 1 for information on Plays, Push 2 for information on movies, Push 3 for information on Museums." If you push 2, the machine may come back, "Push 1 for movies on the south side of town, Push 2 for movies on the

north side of town, etc." That's a very simple interactive voice response (also called Voice Processing) application. There are literally thousands of others. A bank might say "Thanks for calling the XYZ Bank, if you'd like your balance, punch in your account number now." A cable TV company might say "Thanks for calling ABC TV, if you'd like to see tonight's feature movie, push 1. $5.00 will be billed to your account." There are applications for Interactive Voice Response in every industry, in every company. Hint: Think customer! What can we do to make our customers' lives easier? Can we help them find our nearest store? Can we tell them where to get the stuff we sell them repaired? Would they like to know where the order is they placed a week ago?

INTERCEPT SERVICE
A service of the local phone in which a phone call is redirected by an operator or a recording to another phone number or a message. Intercept could also be sold to companies who have moved and would like to retain a recording and/or a phone advertisement for many months after they've moved.

INTERCONNECT COMPANIES
Companies which sell, install and maintain telephone systems for end users, typically businesses.

INTERLATA
Telecommunications services that originate in one and terminate in another Local Access and Transport Area (LATA). Under provisions of Divestiture, the Bell operating companies cannot provide Inter-LATA service, but can provide Intra-LATA service. Some LATAs are very large. So some "local" phone companies provide the equivalent of long distance service. And some of these phone companies have different pricing packages. Some of these packages are cheap, but not highly-publicized. See also LATA.

IVR

Interactive Voice Response.

IXC

InterExchange Channel, or InterExchange Carrier -- as contrasted to the LEC -- the Local Exchange Carrier, a new word for a local phone company. InterExchange Carriers used to be called "Other Common Carriers," except that didn't include AT&T. Now, AT&T, MCI, Sprint and all the Other Common Carriers are called InterExchange Carriers.

L

LAN

Local Area Network. A short distance network (typically within a building or campus) used to link together computers and peripheral devices under some form of standard control. See RING and ETHERNET.

LATA

Local Access and Transport Area; one of 161 local telephone service areas in the United States. As a result of the Bell divestiture that now distinguishes local from long-distance service, switched calls with both endpoints within the LATA (intraLATA) are generally the sole responsibility of the local telephone company, while calls that cross outside the LATA (interLATA) are passed on to an interexchange carrier.

LEC

Local Exchange Company. The local phone companies, which can be either a Bell Operating Company (BOC) or an independent (e.g. GTE) which provides local transmission services. Prior to divestiture, the LECs were called telephone companies or telcos.

LINE POWERED
Telephone equipment that is powered solely by the CO talk battery supplied in a standard phone line.

LOCAL LOOP
The physical wires that run from the subscriber's telephone set, or PBX or key telephone system, to the telephone company central office. Increasingly, the local loop now goes from the main distribution frame in the basement to the phone company. And the subscriber is responsible for getting his/her wires from the box in the basement to his phone, PBX or key system.

LOOP
1. Typically a complete electrical circuit. 2. The loop is also the pair of wires that winds its way from the central office to the telephone set or system at the customer's office, home or factory, i.e. "premises" in telephones. 3. In computer software. A loop repeats a series of instructions many times until some prestated event has happened or until some test has been passed.

LOOP START
You "start" (seize) a phone line or trunk by giving it a supervisory signal. That signal is typically taking your phone off hook. There are two ways you can do that -- ground start or loop start. With loop start, you seize a line by bridging through a resistance the tip and ring (both wires) of your telephone line.

LOOP TIMING
A way of synchronizing a circuit that works by taking a synchronizing clock signal from incoming digital pulses.

LOW LEVEL LANGUAGE
A programming language that uses symbols -- one step away from the machine language of a computer. Low level computer languages such as Assembler and C which actually manipulate the bits in computer registers. Higher level languages such as Basic and Fortran will take care of the piddling details of doing specific functions when you give it a broad command like "PRINT". In a lower level language, you must provide all the details of instruction necessary in the code (program) to perform the operation. It is possible to do this by calling standard routines, but still takes up the programmers' time in deciding which routines, and keeping the registers straight as he designs the program.

MAILBOX
Messages belonging to a single owner in a voice mail system. Today, these will be recorded voice messages, but increasingly mailboxes will include E-mail and fax documents.

MCA
Micro Channel Architecture.

MEDIA PROCESSING
The processing of transactions during a telephone call; these transactions may include fax operations, speech recognition and synthesis, Touch Tone recognition, voice and fax store-and-forward messaging, and the conversion of messages from one format to another (such as from text to voice, or from fax to text).

MENU
Options displayed on a computer terminal screen or spoken by a voice processing system. The user can choose what he wants done by simply

choosing a menu option -- either typing it on the computer keyboard, hitting a touch tone on his phone, or speaking a word or two. There are basically two ways of organizing computer or voice processing software -- menu-driven and non-menu driven. Menu-driven programs are easier for users to use. But they can only present as many options as can be reasonably crammed on a screen or spoken in a few seconds. Non-menu driven screens can allow more any alternatives. But they're much more complex and frightening. It's the difference between receiving a bland "A" or "C" prompt on the screen -- as in MS-DOS and receiving a menu of "Press A if you want Word Processing," "Press B if you want Spread Sheet," etc. It's very easy to write menus in MS-DOS using BATch files. See also AUDIO MENUS.

MICRO CHANNEL

A proprietary 32-bit bus developed by IBM for its PS/2 family of computers' internal expansion cards. Also offered by Tandy and other vendors.

MICROCODE

Programmed instructions that typically are unalterable. Usually synonymous with firmware and programmable read-only-memory (PROM).

MODULATION

The process of varying some characteristic of the electrical carrier wave as the information to be transmitted on that carrier wave varies. Three types of modulation are commonly used for communications, Amplitude Modulation, Frequency Modulation, and Phase Modulation. And there are variations on these themes called Phase Shift Keying (PSK) and Quadrature Amplitude Modulation (QAM).

MSI

Modular Station Interface. A Dialogic board that interfaces analog phones to PEB-based products.

MTBF
Mean Time Between Failure. The length of time a user may reasonably expect a device or system to work before an incapacitating fault occurs.

MTTR
Mean Time to Repair. The average time required to return a failed device or system to service.

MULTIPLEXER
Electronic equipment which allows two or more signals to pass over one communications circuit. That "circuit' may be a phone line, a microwave circuit, a through-the air TV signal. That circuit may be analog or digital. There are many multiplexing techniques to accommodate both.

NETWORK
1. Networks are common in our lives. Think about trains and phones. A networks ties things together. Computer networks connect all types of computers and computer-related things together -- terminals, printers, modems, door entry sensors, temperature monitors, etc. The networks we're most familiar with are long distance ones, like phones and trains. But there are also Local Area Networks (LANs), which exist within a limited geographic area -- like the few hundred feet of a small office, or they can span an entire building, or even touch a "campus," such as a university, or industrial park. There are also Metropolitan Area Networks (MANs). See LAN.

NETWORK BOARD
A board device designed to act as an interface between a compter-based signal processilng system and a telephone network.

NETWORK INTERFACE
The point of inter-connection between Telephone Company communications facilities and terminal equipment, protective apparatus or wiring at a subscriber's premises. The network interface or demarcation point shall be located on the subscriber's side of the Telephone Company's protector, or the equivalent thereof in cases where a protector is not employed, as provided under the local telephone company's reasonable and nondiscriminatory standard operating practices.

NETWORK INTERFACE MODULE
Electronic circuitry connecting a workstation (typically a personal computer) to the telephone network. Network interface modules come in as many flavors as there are ways of connecting to the telephone network -- from simple loop start phone lines to complex primary rate interfaces (PRI) on ISDN. Usually, the network interface module slides into one of the expansion slots inside a personal computer. The card transmits and receives messages from the resource modules connected to the telephone network.

NNX/NXX
A three-digit code to identify the central office in which N is any digit 2 to 9 and X is any digit. Thc first threc digits of a North American telephone number. Originally only NNX codes were used. Now subscriber dials 1+ in making a direct distance dialed toll call and the code may be NXX. Area codes are also being changed to NXX codes.

NODE
An independent SCSA unit in a distributed processing SCSA network, consisting of one or more resource and/or network boards, and one or more SCxbus adapter boards. Communication between nodes take place via the SCxbus. From a device programming point of vicw, a node is simply an addressable system unit which contains boards connected by an SCbus.

NOVELL TELEPHONY SERVICES
Novell's telephony server product, providing a single standard client-server call control interface (inncluding an API for programmers) to many of the diverse PBX systems on the market today.

OAI
Open Application Interface. Basically an opening in a telephone system that lets you link a computer to that phone system and lets the computer command the phone system to answer, delay, switch, hold etc. calls. The term is also called PHI -- as in PBX-Host-Interface. The term OAI was first used by PBX makers, NEC and InteCom. And now the term has become somewhat generic, like all good things. Essentially every manufacturer of phone systems is evolving towards open application interfaces of their own.

OFF-HOOK
When the handset is lifted from its cradle it's Off-Hook. Lifting the hookswitch alerts the central office that the user wants the phone to do something like dial a call. A dial tone is a sign saying "Give me an order." The term "off-hook" originated when the early handsets were actually suspended from a metal hook on the phone. When the handset is removed from its hook or its cradle (in modern phones), it completes the electrical loop, thus signaling the central office that it wishes dial tone. Some leased line channels work by lifting the handset, signaling the central office at the other end which rings the phone at the other end. Some phones have autodialers in them. Lifting the phone signals the phone to dial that one number. An example is a phone without a dial at an airport, which automatically dials the local taxi company. All this by simply lifting the handset at one end -- going "off-hook."

ON-HOOK
When the phone handset is resting in its cradle. The phone is not connected to any particular line. Only the bell is active, i.e. it will ring if a call comes in. See ON-HOOK DIALING and OFF-HOOK.

ON-HOOK DIALING
Allows a caller to dial a call without lifting his handset. After dialing, the caller can listen to the progress of the call over the phone's built-in speaker. When you hear the called person answer, you can pick up the handset and speak or you can talk hands-free in the direction of your phone, if it's a speakerphone. Critical: Many phones have speakers for hands-free listening. Not all phones with speakers are speakerphones -- i.e. have microphones, which allow you to speak, also.

OPEN
An acronym for **O**pen **P**latform **EN**vironment. A term coined by Dialogic to denote its commitment to open standards, formalizing its customers' access to information and improving the overall utility of its various voice processing platforms.

OPEN DEVELOPMENT PROGRAM
Companies belonging to Dialogic's Open Development Program can purchase technical specifications for Dialogic's PEB, AEB, SC, and SCx expansion busses and installable Device Driver Server (DDS). Get a list from Dialogic.

OPEN SOLUTIONS DEVELOPERS
These are companies which build complete turnkey call processing solutions for end users. Get a list from Dialogic.

OPEN TOOLKIT DEVELOPERS
These companies provide applications generators in several operating systems, including MS-DOS, UNIX, OS/2 and Windows. See Chapter 12.

OPERATOR SERVICES
Any of a variety of telephone services which need the assistance of an operator. For example, such services typically include collect calls, third-party billed calls, person-to-person calls.

OUT-OF-BAND SIGNALING
Signaling that is separated from the channel carrying the information -- the voice, data, video, etc. Typically the separation is accomplished by a filter. The signaling includes dialing and other supervisory signals. Out-band-band hardware signaling takes place on its own data path, without the necessity of "bit robbing" or other methods of mixing signals into a data stream.

PABX
Private Automatic Branch Exchange. Originally, PBX was the word for a switch inside a private business (as against one serving the public). PBX means a Private Branch Exchange. Such a "PBX" was typically a manual device, requiring operator assistance to complete a call. Then the PBX went "modern" (i.e. automatic) and no operator was needed any longer to complete outgoing calls. You could dial "9." Thus it became a "PABX." Now all PABXs are modern. And a PABX is now commonly referred to as a "PBX." Some manufacturers have tried to make their PBX appear different by calling it something else. Siemens/Rolm calls theirs the "CBX" (Computerized Branch Exchange). Some others call theirs the "EPABX" (the electronic Private Automatic Branch Exchange. Then there are the special ones, like SRX's SRX, NEC's IMS (Information Management System), etc.

PAM
Pulse Amplitude Modulation. The process of representing a continuous analog signal (a voice conversation) with a series of discrete analog

samples. This concept is based on the information theory which suggests that the signal can be accurately recreated from a sufficient sample. Why bother? Sampling allows several signals to then be combined on a channel that otherwise would only carry one telephone conversation.) PAM was used as part of a method of switching phones calls in several PBXs. It is not a truly "digital" switching system. PAM is the basis of PCM, pulse code modulation. See PCM.

PART 68 REQUIREMENTS (REGISTRATION)
Specifications established by the FCC as the minimum acceptable protection which communications equipment must provide the telephone network. Meeting these requirements does not certify that equipment performs any task. Equipment which is sold for connection to the public network in the US must conform to Part 68.

PBX
Private Branch eXchange. A private (i.e. you, as against the phone company owns it), branch (meaning it is a small phone company central office), exchange (a central office was originally called a public exchange, or simply an exchange). In other words, a PBX is a small version of the phone company's larger central switching office. A PBX is also called a Private Automatic Branch Exchange, though that has now become an obsolete term. In the very old days, you called the operator to make an external call. Then later someone made a phone system that you simply dialed nine (or another digit -- in Europe it's often zero), got a second dial tone and dialed some more digits to dial out, locally or long distance. So, the early name of Private Branch Exchange (which needed an operator) became Private AUTOMATIC Branch Exchange (which didn't need an operator). Now, all PBXs are automatic. And now they're all called PBXs, except overseas where they still have PBXs that are not automatic.

PCM
Pulse Code Modulation. PCM is a method of taking an analog voice signal and encoding it into a digital bit stream. First, the amplitude of the

voice conversation is sampled. This is called PAM, Pulse Amplitude Modulation. This PAM sample is then coded into a binary (digital) number. This digital number consists of zeros and ones. It can then be switched and transmitted digitally. There are three basic advantages to PCM voice. They are the three basic advantages of digital switching and transmission. First, it is less expensive to switch and transmit a digital signal. Second, by making an analog voice signal into a digital signal, you can interleave it with other digital signals -- such as those from computers or facsimile machines. Third, a voice signal which is switched and transmitted end-to-end in a digital format will usually come through "cleaner," i.e. have less noise, than one transmitted and switched in analog. The reason is simple: An electrical signal loses strength over a distance. It must then be amplified. In analog transmission, everything is amplified, including the noise and static the signal has collected along the way. In digital transmission, the signal is "regenerated," i.e. put back together again, by comparing the incoming signal to a logical question: Is it a one or a zero? Then, the signal is regenerated, then it is "amplified and sent along its way.

PCM refers to a technique of digitization. It does not refer to a universally accepted standard of digitizing voice. The most common PCM method is to sample a voice conversation at 8000 times a seconds. The theory is that if the sampling is at least twice the highest frequency on the channel, then the result sounds OK. Thus, the highest frequency on a voice phone line is 4,000 Hertz. So one must sample it at 8,000 times a second. Many PCM digital voice conversations are typically put on one communications channel. In North America, the most typical channel is called the T-1 (also spelled T1). It places 24 voice conversations on two pairs of copper wires (one for receiving and one for transmitting). It contains 8000 frames each of 8 bits of 24 voice channels plus one framing (synchronizing bit) bit which equals 1,544,000 bits per second. i.e. 8000 x (8 x 24 + 1) equals 1.544 megabits.

Europe uses a different scheme for multiplexing voice conversations. It is based not on 24 voice channels, but on 32 channels. This scheme keeps two of the 32 channels for control and actually transmits 30 voice conversations at a data rate of 2,048,000 bits per second. The European system is calculated as 8 bits x 32 channels x 8000 frames per second. European PCM multiplexing is not compatible with North American multiplexing. The two systems cannot be directly connected. Some PBXs in the U.S. conform to the U.S. standard only. Some (very few) conform to both. Both the European and North American T-1 "standards" have now been accepted as ISDN "standards." In addition to PCM, there are many other ways of digitally encoding voice. PCM remains the most common. See E-I, T-1 and VOICE COMPRESSION.

PCM-30

Short name of international 2.048 Mbps T-1 (also known as E1) service derived from the fact that 30 channels are available for 64 Kbps digitized voice each using pulse code modulation (PCM).

PEB

PCM Expansion Bus. Dialogic's name for the digital electrical connection between its network interface modules and its resource modules. This bus is now "open." Technical specifications are available, thus enabling outsiders to create their own resource modules and/or network interface modules. See also AEB, which is Dialogic's Analog Expansion Bus.

PERSONAL IDENTIFICATION NUMBER

PIN number. 1. An AT&T term meaning the last four digits of your AT&T, MCI Bell operating company Credit Card -- the card you use for making long distance numbers. 2. Some banks and financial institutions issue credit cards for machine, teller-less banking. These machines, called Automated Teller Machines, ask you for a password consisting of several numbers or characters. These are not on your credit card. These numbers or characters, called PIN numbers, are designed to make sure the right person is using your card. It's not a good idea to use your birthday as you PIN number.

PHONEMAIL

A ROLM term for Voice Mail. Rolm's PhoneMail is a voice messaging system that provides (1.) telephone answering (with the user's own greeting), (2.) the capability to store and forward voice messages, and (3.) the capability to turn on a message waiting light or message on the recipient's phone. PhoneMail can be used positively -- to speed the flow of information. It can also be used negatively -- to allow the user to "hide behind" the system and avoid the outside world and anyone in the outside world who might actually want to buy something. See also VOICE MAIL.

PIN NUMBER

Personal Identification Number. A group of characters entered as a secret code to gain access to a computer system, such as the one that completes long distance calls. See PERSONAL IDENTIFICATION NUMBER.

PORT

A point of access into a computer or a network. The interface between a computer and its peripheral devices such as printers and modems.

POTS

Plain Old Telephone Service. The basic service supplying standard single line telephones, telephone lines and access to the public switched network. Nothing fancy. No added features. Just receive and place calls. Nothing like Call Waiting or Call Forwarding. They are not POTS services. Pronounced POTS, like in pots and pans.

PREDICTIVE DIALER

See PREDICTIVE DIALING.

PREDICTIVE DIALING
An automated method of making many outbound calls without people and then passing answered calls to a person as the calls are answered. Here's the story: Imagine a bunch of operators having to call a bunch of people. Those calls may be for collections. They may be for employee callups. They may be for alumnae fund raising. When it's done manual, here's how it works: Before each call operators spend time reviewing paper records or computer terminal screens, selecting the person to be called, finding the phone number, dialing the numbers, listening to rings, listening to phone company intercepts, busy signals and answering machines. Operators also spend time updating the records after each call. Predictive dialing automates this process, with the computer choosing the person to be called and dialing the number and only passing it to an operator when a real live human being answers. There are enormous productivity gains made by screening out answering machines, busy signals, network busy signals, non-completed calls, operator intercepts etc. The result is productivity increases of 300% to 600%. According to generally accepted industry lore, a well-run manual dialing center can get its people talking on the phone for 25 minutes an hour. With a predictive dialer you can get them on the phone making sales, collecting money, etc. for 55 minutes an hour. It's a major productivity gain.

True predictive dialing should not be confused with automated dialing. True predictive dialing has complex mathematical algorithms that consider, in real time, the number of available telephone lines, the number of available operators, the probability of getting no answer, a busy signal, a disconnected number, operator intercept or an answering machine, the time between calls required for maximum operator efficiency, the length of an average conversation and the average length of time the operators need to enter the relevant data. Some predictive dialing systems constantly adjust the dialing rate by monitoring changes in all these factors.

Some people don't like the term "predictive dialing," since they think it's getting "a bad rap" in Washington, DC by being associated with junk phone calls. As a result some people would prefer to call it Computer Aided Dialing.

PREVIEW DIALING

Preview dialing is a term used to describe an automatic dialer. Preview dialing is also called "screen dialing" or "cursor dialing." Typically the prospect's account information and/or phone number appears on the screen BEFORE the call is made. Thus the agent can "preview" the number, the screen, the customer. If the agent wants to make the call, the agent hits a key, such as "Enter" and the computer dials the number. In some preview dialing equipments, the agent must hit a key if he/she DOESN'T want the the number dialed. Contrast preview dialing with Predictive Dialing where the computer makes all the dialing decisions and presents the calls to the agent only after they are connected. Predictive dialing is a lot faster than preview dialing. See PREDICTIVE DIALING.

PRI

See PRIMARY RATE INTERFACE.

PRIMARY RATE INTERFACE

PRI. The ISDN equivalent of a T-1 circuit. The Primary Rate Interface (that which is delivered to the customer's premises) provides 23B+D (in North America) or 30B+D (in Europe) running at 1.544 megabits per second and 2.048 megabits per second, respectively. There is another ISDN interface. It's called the Basic Rate Interface. It delivers 2B+D over either one or two pairs. In ISDN, the "B" stands for Bearer, which is 64,000 bits per second, which can carry PCM-digitized voice or data.

PRINTED CIRCUIT BOARD
PCB. Flat material (fiberglass/epoxy) on which electronic components mount. A PCB also provides electrical pathways, called traces, that connect components. Printed circuit boards are what PBXs and computers are made of these days. Be careful when you're replacing PCBs. They're usually very sensitive to static electricity. Handle them only when you're attached to a static electricity strap that is properly grounded. Lay them down only on a surface you're sure is static electricity free. And don't touch the components on PCBs whatever you do.

PRIVATE BRANCH EXCHANGE
PBX. Term used now interchangeably with PABX. See PABX.

PROGRAM LOGIC
The particular sequence of instructions in a program.

PROGRAMMING LANGUAGE
A language used by a programmer to develop instructions for the computer. It is translated into machine language by language software called assemblers, compilers and interpreters. Each programming language has its own grammar and syntax.

PROMPTS
1. Recorded instructions delivered by voice processing units. Prompts may include MENUS or other information that is played each time you get into the system. 2. Messages from the computer instructing the user on how to use the system. See MENU and AUDIO MENU.

PROSODY
Intonation. In text to speech, prosody refers to how natural it sounds -- the ups and downs of the sentence.

PSTN
Public Switched Telephone Network. An abbreviation used by the CCITT.

PULSE AMPLITUDE MODULATION
PAM. A technique for placing binary information on a carrier to transmit that information. PAM is a technique for analog multiplexing. The amplitude of the information being modulated controls the amplitude of the modulated pulses. Samples of each input voltage are placed between voltage samples from other channels. The cycle is repeated fast enough so the sampling rate of any one channel is more than twice the highest frequency transmitted. See also PAM and PCM.

PULSE CODE MODULATION
PCM. The most common and most important method in North America in which a telephone system samples a voice signal and converts that sample into an equivalent digital code. PCM is a digital modulation method that encodes a Pulse Amplitude Modulated (PAM) signal into a PCM signal. See PCM and T-1.

PULSE DENSITY
In T-1, since "O"s are represented by no pulse and "1"s by alternating pulses, pulse density refers to the number of no pulse ("O") periods allowed before a pulse ("1") must occur. Typically, no more than 15 no pulse periods ("O"s) are allowed before a pulse ("1") must occur.

PULSE DIALING
One or two types of dialing that uses rotary pulses to generate the telephone number.

PULSE DISPERSION
The spreading out of pulses as they travel along an optical fiber.

PULSE DURATION MODULATION

PDM. That form of modulation in which the duration of the pulse is varied in accordance with some characteristic of the modulating signal.

PULSE LINK REPEATER

A signaling set that interconnects the E and M leads of two circuits. In E & M signaling, a device that interfaces the signal paths of concatenated trunk circuits. Such a device responds to a ground on the "E" lead of one trunk by applying -48Vdc to the "M" lead of the connecting trunk, and vice versa. This function is a built-in, switch-selectable option in some commercially available carrier channel units.

PULSE OVERSHOOT

In T-1, the amount of signal voltage that can remain at the trailing end of a pulse. It can be no more than 10-30% of the pulse amplitude. Also called afterkick.

PULSE-POSITION MODULATION

PPM. That form of modulation in which the positions in time of the pulses are varied in accordance with some characteristic of the modulating signals, without modifiying the pulse width.

PULSE REPETITION FREQUENCY

PRF. In radar, the number of pulses that occur each second. Not to be confused with transmission frequency which is determined by the rate at which cycles are repeated within the transmitted pulse.

PULSE STUFFING

When timing signals on digital circuits get out of whack, some method of allowing mismatches must be provided. In time division multiplexing, this is called pulse stuffing. One stream of data has bits added to it so its final rate is the same as the master clock.

PULSE TRAIN
The resulting electronic impulses that transmit encoded information.

PULSE WIDTH
In T-1, refers to the width (at half amplitude) of the bipolar pulse (typically 324 + or -45 nsec).

QUANTIZE
The process of encoding a PAM signal (Pulse Amplitude Signal) into a PCM signal (Pulse Code Modulation). See QUANTIZING and QUANTIZATION.

QUANTIZING
The second stage of pulse code modulation (PCM). The waveform samples obtained from each communication channel are measured to obtain a discrete value of amplitude. These quantized values are converted to a binary code and transmitted to a distant location to reconstruct the original waveform. See QUANTIZATION.

QUANTIZATION
Take an analog voice signal, sample it and put numbers on those samples. That's called quantization. Here's an explanation from Understanding Telephone Electronics: "A circuit called a quantizer takes in the analog signal and produces an equivalent number. Threshold levels are established and numbers are assigned to the analog samples as their amplitudes fall within the bands formed by the threshold numbers. (For example, everything within the range 1000.05 and 1000.06 Hz is given the number 14568.) The assigned number in most costs is an approximation rather than

a true value because the true value would require many more bits (i.e. more threshold limits) in the binary code...This approximation causes an error which is the difference between the approximate number and the true sample. This quantization error adds noise to the signal, called quantization noise, which is heard in the telephone as hissing. Quantization noise can be reduced by making the threshold bands narrower...However, providing more intervals requires more bits in the binary code. Therefore, more bandwidth is needed. There is a tradeoff between small quantizing intervals (higher bandwidth, lower noise), and fewer intervals (lower bandwidth, higher noise).

QUANTIZATION NOISE
Signal errors which result from the process of digitizing (and therefore ascribing finite quantities to) a continuously variable signal. See QUANTIZATION.

REAL TIME
A voice telephone conversation is conducted in Real Time. That is, there is no perceived delay in the transmission of the voice message or in the response to it. This concept often applies to interaction between a computer and a terminal. See also REAL TIME CAPACITY. In data processing or data communications, real time means the data is processed the moment it enters a computer, as opposed to BATCH processing where the information enters the system, is stored and is operated on a later time.

REGISTRATION NUMBER (FCC PART 68)
Approval number given to telephone equipment to certify that a particular device passes the tests defined in Part 68 of the FCC Rules. These tests

certify the phone won't cause any harm to the public network. They do not attest to the commercial value of the product, nor whether it will (or won't) sell. See PART 68.

REMOTE DIAGNOSTICS
You own a phone system. You have a service company. There's some problem with it. Instead of sending a technician out, your service company dials your PBX from a data terminal or PC and "asks" your PBX in computerese what's wrong with it. If it isn't too broken, it will come back and give you some indication. This is called remote diagnostics.

REMOTE PROGRAMMING
Dial your phone system with your friendly personal computer, modem and a communications software package and you can change the telephone system's programming remotely. This feature is great for companies with telephone systems in many locations. They can all be run from one central point. This feature is also great. You may want some changes made on your system. It's obviously a lot cheaper for your vendor to make those changes from his office than having to visit yours. It's also a lot faster. See REMOTE DIAGNOSTICS.

RESOURCE MODULE
A Dialogic term referring to devices which perform specific voice processing functions, such as voice compression, voice recognition, facsimile transmission and reception and conversion of computer text to spoken words over the phone. Resource modules are typically connected to the telephone network through devices known as "network interface modules."

RING
1. As in Tip and Ring. One of the two wires (the two are Tip and Ring) needed to set up a telephone connection. 2. Also a reference to the ringing

of the telephone set. 3. The design of a Local Area Network (LAN) in which the wiring loops from one workstation to another, forming a circle (thus, the term "ring"). In a ring LAN, data is sent from workstation to workstation around the loop in the same direction. Each workstation (which is usually a PC) acts as a repeater by re-sending messages to the next PC in the ring. The more PC's, the slower the LAN. Network control is distributed in a ring network. Since the message passes through each PC, loss of one PC may disable the entire network. However, most ring LANs have a way of recovering very fast should one PC die or be turned off. If it dies, you can remove it physically from the network. If it's off, the network senses that and the token ignores that machine. In some token LANs, the LAN will close around a dead workstation and join the two workstations on either side together. If you lose the PC doing the control functions, another PC will jump in and take over. This is how the IBM Token- Passing Ring works. See TOPOLOGY and RING.

RINGING VOLTAGE
In addition to talk battery, a Central Office provides ringing signaling. Ring Voltage is generally 70 to 90 volts at 17 Hz to 20 Hz.

RJ-11
RJ-11 is is a six conductor modular jack that is typically wired for four conductors (i.e. four wires). Occasionally it is wired for only two conductors -- especially if you're only wiring up for tip and ring. The RJ-11 jack (also called plug) is the most common telephone jack in the world. The RJ-11 is typically used for connecting telephone instruments, modems and fax machines to a female RJ-22 jack on the wall or in the floor. That jack in turn is connected to twisted wire coming in from "the network" -- which might be a PBX or the local telephone company central office. RJ-22 wiring is typically flat. None of its conductors (i.e. wires) are twisted. You cannot use flat cable for high-speed data communications, like local area networks. See also RJ-22 and RJ-45.

RJ-14

A jack that looks and is exactly like the standard RJ-11 that you see on every single line telephone. Whereas the RJ-11 defines one line -- with the two center, red and green, conductors being tip and ring, the RJ-14 defines two phone lines. One of the lines is the "normal" RJ-11 line -- the red and green center conductors. The second line

RJ-45

The RJ-45 is the 8-pin connector used for data transmission over standard telephone wire. That wire could be flat or twisted. And it's very important that you know what you're working with. You can easily use flat wire for serial data communications up to 19.2 Kbps. Up to that speed you're connecting with your wire to a dataPBX, a modem, a printer or a printer buffer. If you wish to connect to a 10BaseT local area network, which you also do with a RJ-45, you must use twisted wire. You can typically tell the difference by looking at the cable. If it's flat grey satin (like a typical phone wire, only bigger) than it's probably untwisted. If it's circular, then it's probably twisted and therefore good for LANs. RJ-45 connectors come into two varieties -- keyed and non-keyed. Keyed means that the male RJ-45 plug has a small, square bump on its end and the female RJ-45 plug is shaped to accommodate the plug. A keyed RJ-45 plug will not fit into a female, non-keyed (i.e. normal) RJ-45. See RJ-11 and RJ-22.

ROBBED-BIT SIGNALING

This explanation from Gary Maier of Dianatel: ISDN is the key to future sophisticated telephone network services with its dynamic, highly configurable T-1 connection (also called PRI connection). Since T-1 is a common method of carrying 24 telephone circuits, many wonder about the uses for ISDN, especially when they learn ISDN signaling requires an entire voice channel, reducing today's T-1 from 24 voice channels to 23. But the popular signaling mechanism of "robbed bit" signaling in T-1 has serious limititations. Robbed bit signaling typically uses bits known as the A and B bits. These bits are sent by each side of a T-1 termination and are buried in the voice data of each voice channel in the T-1 circuit. Hence the term "robbed bit" as the bits are stolen from the voice data. Since the bits are stolen so infrequently, the voice quality is not compromised by much. But the available signaling combinations are limited to ringing, hang up, wink, and pulse digit dialing. In fact, the limitations are obvious when one recognizes DNIS and ANI information are sent as DTMF tones.

This introduces a problem: time. Each DTMF tone requires at least 100 milliseconds to send, which in a DNIS and ANI situation with 20 DTMFs will take at least two full seconds. There is also a margin for error in transmission or detection, resulting in DNIS or ANI failures. With the explosion of telephone related services, the telephone companies are turning to ISDN PRI to provide the more complicated and exact signaling required for new services. ISDN employs a more robust method of signaling. ISDN uses a T-1 circuit as 23 voice channels and one signaling channel. The term 23B plus D refers to 23 bearer (voice) channels and 1 Data (signaling) channel. The data channel carries the signaling information at a rate of 64 kilobits per second. This speed is many times greater than some of the most powerful modems available. Because of this high speed, telephone calls can be placed more quickly, and because of the protocol used, DNIS or ANI transmission failures are impossible.

Additionally, since no bits are "robbed" from the voice channels, the voice quality is better than that of Robbed Bit signaling on today's T-1 circuits. Also, computer modems and high speed faxes can use the voice channel for sending digital data instead of the traditional analog bit "noise." Therefore, ISDN PRI offers the end user countless new service capabilities. One channel could be used for faxing, another for modem data, several for video, another for a LAN and the remainder for voice. Suddenly, the average T-1 circuit becomes a pipeline for all communications! Increasingly long distance carriers are using ISDN PRI to provide inbound 800 calls with ANI and DNIS and re-routing skills. Dianatel makes some of the most sophisticated ISDN interface equipment around.

ROTARY DIAL
The circular telephone dial. As it returns to its normal position (after being turned) it opens and closes the electrical loop sent by the central office. Thus it generates pulses for each digit dialed. You can hear the "clicks". The number "seven," for example consists of seven "opens and closes," or seven clicks. You can dial on a rotary phone without using the rotary dial. Simply depress the switch hook quickly, allowing pauses in between to signify that you're about to send a new digit. It's a good party trick.

S

SAMPLING RATE
The number of times per second that an analog signal is measured and converted to a binary number -- the purpose being to convert the analog signal to a digital analog. The most common digital signal -- PCM -- samples 8,000 times a minute.

SCSA
Signal Computing System Architecuture. A new standard for the telecom/voice processing/computing industry announced by Dialogic in early Spring 1993. According to TELECONNECT Magazine, this SCSA standard is remarkable for several things:

1. On the day of its announcement over 70 telecom and voice processing companies publicly endorsed SCSA, committed to work with it and are clearly planning to work it.

2. With SCSA -- a standard for PC/LANs and VME-backplaned computers -- you can build much larger telecom switches and much larger call and voice processing boxes. Previous standards, like AEB, PEB and MVIP, were basically limited to what you could do with one PC. Now PCs can be joined together. With SCSA, you can put 16 T-1 lines, or 512 voice lines in one PC and join together 16 PCs, for a total of 16 x 16 x 24 = 6,144 lines! That's a central office built out of networked PCs. A mainframe built out of a LAN. The SCSA joining is not via LAN or LAN-emulation. That would be too slow and the transmission too bursty (great for data, lousy for voice). It's via an SCbus -- something that looks and works like a PBX backplane.

3. SCSA incorporates virtually every other standard in PC-based swiching -- including the most popular ones, Mitel's ST-Bus, MVIP, Siemens PCM Highway, AEB and PEB.

4. It's a lot faster and more reliable. All signaling is out of band. There's clock fall-back and time slot bundling. It's more modular, meaning you can start with one PC and grow one at a time. That makes it more "modular" (scalable is the new word). It's also hot pluggable. You don't have to turn off to upgrade.

5. It has applications portability. Tandem, the highly-successful fault-tolerant mini-computer maker, has an SCSA application in a call center. They call it the Tandem Non-Step Call Center. It uses the Tandem 2400 VRU and the 4800 VRU.

SCSA is open, truly open. All its specs and all levels of its specs are available. To that extent, SCSA represents a remarkable gamble by its creator, Dialogic, a telecom/voice processing hardware company. It is encouraging competing manufacturers to build hardware to its specs and gambling that it won't be left in the dust, as IBM was with its PC. (Compaq, not IBM, built the first '386 PC.)

SCSA, as an idea, is revolutionary (for telecom). No one in telecom has ever promolgated an open standard everyone can adopt -- hardware and software vendors. Dialogic has done it to create opportunities by providing great economies of scale for the developers. Write one application, create one applications generator, design one piece of hardware. Erector set telecom/voice processing! Build small. Build large. Just join the bits and pieces together.

SERVICE CONTROL POINTS
SCPs. The local versions of the national SMS/800 number database. SCPs contain the intelligence to screen the full ten digits of an 800 number and route calls to the appropriate, customer-designated long distance carrier.

SIDETONE
A part of the design of a telephone handset which allows you hear your own voice while talking on the telephone. The idea of doing this is to give you a little feedback that the telephone you're speaking on is working. Too much sidetone becomes an echo and is bad. Too little sidetone makes the channel unerring.

SIGNALING
Pertains to the transmission of electrical signals to and from the user's premises and the telephone company central office. Examples of central office signals to the user's premises are ringing (audible alerting) signals, dial tone, speech signals, etc. Signals from the user's telephone include off-hook (request for service), dialing (network control signaling), speech to the distant party, on-hook (disconnect signal), etc.

SIGNAL-TO-NOISE RATIO
The ratio of the usable signal being transmitted to the noise or undesired signal. Usually expressed in decibels. This ratio is a measure of the quality of a transmission.

SIR
Speaker Independent Recognition. See SPEAKER INDEPENDENT RECOGNITION.

SIT TONES
1. Standard Information Tones. These are tones sent out by a central office to a pay phone to indicate that the dialed call has been answered by the distant phone, etc. **2.** Special Information Tones. These are tones for identifying network provided announcements. Here's Bellcore's explanation: Automated detection devices cannot distinguish recorded voice from live voice answer unless a machine-detectable signal is included with the recorded announcement. The CCITT, which specifies signals that may be applied to international circuits, has defined Special Information Tones for identifying network provided announcement. The SIT used to precede machine-generated announcements also alerts the calling customer that a machine-generated announcement follows. Since SIT consists of a sequence of three precisely defined tones, SIT can be machine-detected, and therefore machine-generated announcements preceded by a SIT can be classified. At least four SIT encodings have been defined: Vacant Code (VC), Intercept (IC), Reorder (RO) and No Circuit (NC). With the exception of some small Stored Program Control Systems (SPCSs) and some customer negotiated announcements, Bell operating companies in North America now precede appropriate announcements with encoded SITs to detect and classify announcements.

SMS/800
The national database Service Management System that will rctain all 800 records. This database provides long distance carriers a single interface for 800 number reservations and record maintenance. Developed by Bellcore, the database has been in use by various Regional

Bell Operating Companies (RBOCs) since 1988. The FCC has mandated that a neutral third party administer the database.

SOFTWARE SUPERVISION

"Answer Supervision" is knowing when the person at the other end answers the phone. The main reason for wanting to know this is so that a phone company can start billing the call. There are two ways of doing answer supervision. You can get it from the nation's phone system, i.e. the distant office signals back across the country when the called person picks up the phone. Or you can fake it with "software supervision." Essentially this means there's electronics which "listens" to the call. If it "hears" voice or something like voice, it assumes the conversation has started and it's time to start billing the call. Software supervision is not accurate. But when you haven't got access to real answer supervision (for whatever reason) it's better than the previous alternative, which was "timeout." In timeout answer supervision, the carrier simply assumed the call had begun after a certain number of seconds -- like 30 -- had elapsed with the calling person hanging up. This meant, for example, if you called Grandma and she wasn't there, but you left it ringing, 'cause you knew she took time to answer the phone, then you'd be charged for the call -- even though she didn't answer phone! See also ANSWER SUPERVISION.

SOURCE CODE

A computer program which can be translated by another program into a program that will run on a particular computer. The program which finally runs on that computer is known as the object code.

SPAN

1. Refers to that portion of a high speed digital system than connects a C.O. (Central Office) to C.O. or terminal office to terminal office. **2.** Also called a T-Span Line. A repeatered outside plant four-wire, two twisted-pair transmission line.

SPAN LINE
A T-1 link.

SPEAKER INDEPENDENT (VOICE) RECOGNITION
SIR. Technologies for the automated conversion of speech to accurate and meaningful textual information (typically ASCII). SIR is typically used to accept input from callers to voice processors where the callers are using rotary dial phones instead of touchtone -- Dual Tone Multi-Frequency (DTMF) -- phones. SIR can substitute for the numbers on the DTMF keypad and can add the benefit of a few basic voice commands, e.g., Yes, No, Help, etc. Because computer processing demands are formidable with speaker independent recognition, accurate speaker independent products are created with limited vocabularies. In contrast, trainable or speaker dependent recognizers can feature larger vocabularies at lower prices. SIR has been slowly gaining cceptance in telephone applications. SIR is increasingly used in automated perator assistance applications. SIR will see increased use as system builders respond to pressures to provide voice processing functions to the enormous rotary phone installed base domestically and abroad.

SPEECH CONCATENATION
A term used in voice processing for economical digitized speech playback that uses independently recorded files of phrases or file segments linked together under application program control to produce a customized response in natural sounding language. For example, order status, bank balances, bus schedules or lottery results, etc. Concatenation is done for speed and economy. It lends itself to limited and structured vocabularies that are best stored in RAM (Random Access Memory) or speedily accessible from disk. Concatenation does not replace Text-To-Speech (TTS) as a method of getting the voice processer to deliver its responsese. Concatenation, however, can be an excellent complement to TTS when a voice application demands broad, real time vocabulary production. See TEXT-TO-SPEECH.

SPEECH SYNTHESIS
See TEXT-TO-SPEECH.

SPRINGBOARD
Dialogic's term for a general purpose computing engine designed for a wide range of powerful voice processing applications. It contains two Motorola 56001 DSPs (Digital Signal Processors) and one Intel 80286 12.5 MHz microprocessor. It allows Dialogic's Technology Partners to extend D/12x functions.

STATE MACHINE PROGRAMMING
To control multiple telephone lines in a single voice processing program, a new program structure is required. Dialogic calls this technique state machine programming. Computer Science called state machines "Deterministic Finite State Automata."

STATION
A dumb word for a telephone. Also called an instrument, or a telephone instrument. An extension station is one connected "behind" a PBX or key system. In other words, the PBX or key system is between the station and the telephone central office. We tried to remove the word "station" from this dictionary, but failed. We suspect the word comes from the very old days when the telephone industry was regulated by the Interstate Commerce Commission, (the ICC) which also regulated the railroad industry.

STATION ADAPTERS
Cables and interface assemblies for connecting Dialogic network interface and switching products to telephones or analog telephone lines.

STATION SET
Another word for a common desk telephone.

STORE AND FORWARD
In communications systems when a message is transmitted to some intermediate relay point and stored temporarily. Later the message is sent the rest of the way. Not very convenient for voice conversations, but useful for telex type, and other one way transmissions of messages. Telephone answering machines, as well as voice mailboxes are considered forms of Store and Forward message switching.

SUPERVISED TRANSFER
A call transfer made by an automatic device such as voice response unit which attempts to determine the result of the transfer -- answered, busy, ring no answer -- by analyzing call progress tones on the time.

SUPERVISION
Supervision of a phone call is detecting when a called party has picked up his phone and when that party has hung up. Supervision is used primarily for billing purposes. Not all long distance carriers have supervision capability. It depends on how "equal accessed" they have chosen to be. See ANSWER SUPERVISION and SOFTWARE SUPERVISION.

SWITCH
A mechanical, electrical or electronic device which opens or closes circuits, completes or breaks an electrical path, or selects paths or circuits.

SWITCHHOOK
A switchhook was originally an electrical "switch" connected to the "hook" on which the handset (or receiver) was placed when the telephone was not in use.

SWITCH HOOK FLASH
A signaling technique whereby the signal is originated by momentarily depressing the switch hook.

SYNTHESIZED VOICE
Human speech approximated by a computer device that concatenates basic speech parts (or phonemes) together.

T

T-1
Also spelled T1. A digital transmission link with a capacity of 1.544 Mbps (1,544,000 bits per second). T1 uses two pairs of normal twisted wires, the same as you'd find in your house. T1 normally can handle 24 voice conversations, each one digitized at 64 Kbps. But, with more advanced digital voice encoding techniques, it can handle more voice channels. T1 is a standard for digital transmission in North America. It is usually provided by the phone company and used for connecting networks across remote distances. Bridges and routers are used to connect LANs over T1 networks. There are faster services available, such as T2 and T4, but these are not used much. T1 links can often be connected directly to new PBXs and many new forms of short-haul transmission, such as short-haul microwave systems. It is not compatible with T1 outside the United States and Canada. In Europe T-1 is called E-1 or E1.

Outside of the United States and Canada, the "T-1" line bit rate is usually 2,048,000 bits per second. Japan, France and West Germany impose slight variations that make their formats unique. Only one element remains constant -- the DS-0. The 64 kilobit per channel is universal. Most often it represents a PCM voice signal sampled at 8,000 times per second. However, the form of PCM encoding differs between T-1 (mu-law) and E-1 (A-law companding). According to Bill Flanagan's book, the differences are not so great that a multiplexer cannot convert between them. Conversion of E-1 to T-1 involves both the compression law and the signaling format.

At the higher rate of 2,048,000, 32 time slots are defined at the CEPT interface, but two are used for signaling and other housekeeping chores. Typically 30 channels are left for user information -- voice, video, data, etc. CEPT is the Conference of European Postal and Telecommunications administrations. Standards-setting body whose membership includes European Post, Telephone, and Telegraphy Authorities (PTTs).

For a full explanation of T1 see Bill Flanagan's book The Guide to T-1 Networking. (Call 1-800-LIBRARY for your copy.)

TABLE-DRIVEN
Describing a logical computer process, widespread in the operation of communications devices and networks, where a user-entered variable (e.g. a password) is matched against an array of predefined values (a table of acceptable passwords) and checked for access, authorization, etc. A frequently used process in least cost routing, in network routing, in access security, and in modem operation.

TALK BATTERY
The DC voltage supplied by the central office to the subscriber's loop to operate the carbon transmitter in the handset.

TALK PATH
The tip and ring conductors of a telephone circuit.

TALK-OFF
Talk-off is one hazard of in-band signaling. Talk-off occurs when your voice has enough 2600 Hz energy to activate the 2600 Hz tone- detecting circuits in the central office. The 2600 Hz tone is used for in-band signaling.

TAPI
See Windows Telephony API

TECHNOLOGY DEVELOPERS
Dialogic term for companies which make specialized voice processing devices which can be used with Dialogic products to build end user products. Technology partners include those with skills in facsimile, voice recognition and text-to-speech. Their products will work with Dialogic products.

TELCO
The local telephone company. Often a term of endearment.

TELECOMPUTING
Telephony services administered by a computer. These services include call control and media processing operations.

TELECOMPUTING SERVER
A server software program that allows client applications to take advantage off call control and media processing hardware remotely. See AppServer for an example.

TELEMARKETING
Marketing and sales conducted via the telephone. There are two sides to telemarketing -- incoming and outgoing. Incoming telemarketing is largely run through 800 toll-free IN-WATS numbers and local FX (foreign exchange) lines. Outgoing telemarketing is organized over OUT-WATS lines. Everything from computers to clothes is now sold over the phone. An expanding range of telecom gadgetry is being developed to automate telemarketing -- including automated outbound dialers, voice processing technology, and automatic call distributors. The tone recognition, voice detection and transaction Audiotex and transaction processing capabilities of voice processing gear can be used to enhance all telemarketing applications.

TELEPHONE MANAGER
Apple's telephony API for the Macintosh environment.

TELEPHONY DATABASE TRANSACTIONS
The use of telephony terminal equipment (phone and/or fax) as input/output devices to operate on a computer database.

TELEPHONY SERVER
A server software program that allows client applications to take advantage of call control hardware remotely. See Novell Telephony Server for an example.

TELEPHONY WORKGROUP
A group of people with a common telecomputing requirement: for example, a technical support group might need an application that automatically links incoming calls to client records on a database.

TERMINAL EMULATION
A microcomputer or personal computer mimics, pretends to be (i.e. emulates) a data terminal. It does this with special printed circuit boards inserted into its motherboard and/or special software. For example, TELECONNECT uses the communications software program called Crosstalk to emulate a DEC VT-100, a Digital Equipment Corporation VT-100 terminal. TELECONNECT does this because emulating a DEC VT-100 works better with certain software programs that are called-up remotely.

TEXT-TO-SPEECH (SPEECH SYNTHESIS)
TTS. Technologies for converting textual (ASCII) information into synthetic speech output. This technology is used in voice processing applications that require the production of broad, unrelated and unpredictable vocabularies, e.g., products in a catalog, names and addresses, etc. This technology is appropriately used when system design constraints prevent the more efficient use of speech concatenation alone. See SPEECH CONCATENATION.

TIP & RING

An old fashioned way of saying "plus" and "minus," or ground and positive in electrical circuits. Tip and Ring are telephony terms. They derive their names from the operator's cordboard plug. The tip wire was connected to the tip of the plug, and the ring wire was connected to the slip ring around the jack. A third conductor on some jacks was called the sleeve. That's it. Nothing more sinister. Nothing more interesting.

TONE DIAL

A pushbutton telephone dial that makes a different sound (in fact, a combination of two tones) for each number pushed. The correct name for tone dial is "Dual Tone MultiFrequency" (DTMF). This is because each button generates two tones, one from a "high" group of frequencies -- 1209, 1136, 1477 and 1633 Hz -- and one from a "low" group of frequencies -- 697, 770, 852 and 841 Hz. The frequencies and the keyboard, or tone dial, layout have been internationally standardized, but the tolerances on individual frequencies vary between countries. This makes it more difficult to take a touchtone phone overseas than a rotary phone.

You can "dial" a number faster on a tone dial than on a rotary dial, but you make more mistakes on a tone dial and have to redial more often. Some people actually find rotary dials to be, on average, faster for them. The design of all tone dials is stupid. Deliberately so. They were deliberately designed to be the exact opposite (i.e. upside down) of the standard calculator pad, now incorporated into virtually all computer keyboards. The reason for the dumb phone design was to slow the user's dialing down to the speed Bell central offices of early touch tone vintage could take. Today, central offices can accept tone dialing at high speed. But sadly, no one in North America makes a phone with a sensible, calculator pad or computer keyboard dial. On some telephone/computer workstations you can dial using the calculator pad on the keyboard. This is a breakthrough. It a lot faster to use this pad. The keys are larger, more sensibly laid out and can actually be touch-typed (like touch-typing on a keyboard.) Nobody, but nobody can "touch-type" a conventional telephone tone pad. A tone dial on a telephone can provide access to various special interactive voice response services and features -- from ordering your groceries over the phone to inquiring into the prices of your (hopefully) rising stocks.

TOOLKIT
A Dialogic word for an Applications Generator.

TOOLKIT MEMBERS
An company which has written one or more applications generators for use with Dialogic products. Get a list from Dialogic.

TOUCHTONE
A trademark owned by AT&T for Tone Dialing.

TTS
Text To Speech. A term used in voice processing. See TEXT-TO-SPEECH.

TRUNK
A telephone communication path, or channel, between two points, one of them usually being a telephone company central office or switching center.

T-SPAN
A telephone circuit or cable through which a T-carrier runs. See T-1.

TWISTED PAIR
Two insulated copper wires twisted around each other to reduce induction (thus interference) from one wire to the other. The twists, or lays, are varied in length to reduce the potential for signal interference between pairs. Several sets of twisted pair wires may be enclosed in a single cable. In cables greater than 25 pairs, the twisted pairs are grouped and bound together in a common cable sheath. Twisted pair cable is the most common type of transmission media. It is the normal

cabling from a central office to your home or office, or from your PBX to your office phone. Twisted pair wiring comes in various thicknesses. As a general rule, the thicker the cable is, the better the quality of the conversation and the longer cable can be and still get acceptable conversation quality. However, the thicker it is, the more it costs.

TWO-WIRE CIRCUIT
A transmission circuit composed of two wires -- signal and ground -- used to both send and receive information. In contrast, a four wire circuit consists of two pairs. One pair is used to send. One pair is used to receive.

All trunk circuits -- long distance circuits -- are four wire. A four wire circuit costs more but delivers better reception. All local loop circuits -- those coming from a Class 5 central office to the subscriber's phone system -- are two wire, unless you ask for a four-wire circuit and pay a little more.

UNIFIED MESSAGING
A telecomputing application that treats incoming and outgoing electronic mail, voice messages, and fax messages together. A mixed-media messaging control mechanism.

UNIX
An immensely powerful and complex operating system for computers. UNIX is developed, marketed, championed and trademarked by AT&T. UNIX is very powerful and very complex to use. Because of its power it needs a computer with a large amount of RAM memory. It has been extensively used in universities, and is only now emerging onto the business scene. UNIX allows a computer to handle multiple users and multiple programs simultaneously. It works on many different

computers. This means you can often take applications software which runs on UNIX and often move it -- with little changing -- to a bigger, different computer, or to a smaller, different computer. This process of moving programs to other computers is known as "porting." UNIX was developed in 1969 by Ken Thompson of AT&T Bell Laboratories.

VARI-A-BILL
A new 900 service of AT&T whereby the call's price varies depending on certain events -- the caller punching out some tones on his phone, or a service technician coming on line, etc. Telephone Response Technologies (TRT) supports this feature with their call processing module called Pro/ISDN.

VISUAL VOICE MAIL
An application displaying and controlling voice messages on a desktop computer. Usually associated with unified messaging.

VOICE BOARD
Also called a voice card or speech card. A Voice Board is an IBM PC- or AT-compatible expansion card which can perform voice processing functions. A voice board has several important characteristics: It has a computer bus connection. It has a telephone line interface. It typically has a voice bus connection. And it supports one of several operating systems, e.g. MS-DOS, UNIX. At a minimum, a voice board will usually include support for going on and off-hook (answering, initiating and terminating a call); notification of call termination (hang-up detection); sending flash hook; and dialing digits (touchtone and rotary). See VRU.

VOICE CIRCUIT
The typical analog telephone channel coming into your house or office. It has a bandwidth between 300 Hz and 3000 Hz. This is not sufficient for

high fidelity voice transmission. You'd probably need at least 10,000 Hz. But it's sufficient to recognize and understand the person on the other end.

VOICE DIGITIZATION
The conversion of an analog voice signal into binary (digital) bits for storage or transmission.

VOICE DRIVER
A Dialogic product that comes for MS-DOS, OS/2 and UNIX. In MS-DOS, it is a terminate and stay resident (TSR) program which acts as a central server for MS-DOS based applications. It provides all of the services required to support installable device drivers for each hardware component and for the application. See also DEVICE DRIVER.

VOICE FREQUENCY
VF. An audio frequency in the range essential for transmission of speech. Typically from about 300 Hz to 3000 Hz.

VOICE MAIL
A specialized form of Interactive Voice Response. Includes Voice Messaging and Automated Attendant systems. Denotes customer premise or central office systems that are aimed at automating and enhancing message taking and ATTENDANT console styled services. See VOICE MAIL SYSTEM.

VOICE MAIL SYSTEM
A device to record, store and retrieve voice messages. There are two types of voice mail devices -- those which are "stand alone" and those which profess some integration with the user's phone system. A stand alone voice mail is not dissimilar to a collection of single person answering machines, with several added features; you can instruct the machines (voice mail boxes) to forward messages amongst themselves; you can organize to allocate your friends and business acquaintances their own mail boxes so they can dial, leave messages, pick up messages

from you, pass messages to you, etc; you can also edit messages, add comments and deliver messages to a mailbox at a pre-arranged time. In addition, messages can be tagged "urgent" or "non-urgent" or stored for future listening. The range of voice mail options varies among manufacturers. An integrated voice mail system includes these features. First, it will tell you if you have any messages. It does this by lighting a light on your phone, and/or putting a message on your phone's alpha-numeric display. Second, if your phone rings for a certain number of rings (which number you set), the phone will transfer your caller automatically to your voice mail box, which will answer the phone, deliver a little "I am away" message and then receive and record the caller's message.

VOICE MESSAGE SERVICE
A service typically over dial up phone lines which provides the ability for a phone user to access a voice mail system and leave a message for a particular phone user. See VOICE MAIL SYSTEM.

VOICE MESSAGING
Recording, storing, playing back and distributing phone messages. New York Telephone has an interesting way of looking at voice messaging. NYTel sees it as four distinct areas: **1.** Voice Mail, where messages can be retrieved and played back at any time from a user's "voice mailbox"; **2.** Call Answering, which routes calls made to a busy/no answer extension into a voice mailbox; **3.** Call Processing, which lets callers route themselves among destinations via their touch-tone phones; and 4. Information Mailbox, which stores general recorded information for callers to hear.

VOICE PROCESSING
Another term for INTERACTIVE VOICE RESPONSE.

VOICE RECOGNITION
The ability of a machine (obviously a computer) to understand human speech.

VOICE RESPONSE UNIT
VRU. Another term for INTERACTIVE VOICE RESPONSE. Think of a Voice Response Unit as a voice computer. Where a computer has a keyboard for entering information, an IVR uses remote touchtone telephones. Where a computer has a screen for showing the results, an IVR uses a digitized synthesized voice to "read" the screen to the distant caller. An IVR can do whatever a computer can, from looking up train timetables to moving calls around an automatic call distributor (ACD). The only limitation on an IVR is that you can't present as many alternatives on a phone as you can on a screen. The caller's brain simply won't remember more than a few. With IVR, you have to present the menus in smaller chunks. See TELECOMPUTING; VOICE BOARD.

VOICE STORE AND FORWARD
Voice mail. A PBX service that allows voice messages to be stored digitally in secondary storage and retrieved remotely by dialing access and identification codes. See VOICE MAIL SYSTEM.

VOICE SWITCHED
A device which responds to voice. When the device hears a voice, it turns on and transmits it, while muting the receive side. The most common voice-switched device is the common desk speakerphone. With voice switching, it's easy to hog a circuit. Just keep making a noise. Watch out for voice hogging. If you're calling someone and waiting for them by listening in on your speakerphone, mute your speakerphone. This way you'll hear them when they answer.

VS&F
Voice Store and Forward. Voice is digitally encoded, sent to large storage devices and later forwarded to the recipient. See VOICE MAIL.

WATS
Wide Area Telecommunications Service. WATS is basically discounted toll service. It is provided by all long distance and local phone companies. AT&T started WATS but forgot to trademark the name, so now every supplier uses it as a generic name. There are two types of WATS services -- in and out WATS, i.e. those WATS lines that allow you to dial out and those which allow you to receive incoming calls (the typical 800 line service). You subscribe to in and out-WATS services separately. In the old days you needed separate in and out lines to handle the in and out WATS services. But these days you can choose to have in and out WATS on the same line. This is not particularly brilliant traffic engineering, since you can't receive an incoming 800 call if you're making an outgoing call.

WINDOWS TELEPHONY API
Microsoft's standard call control programming interface for the Windows environment. Microsoft and Intel introduced Windows Telephony as a concept in the Spring of 1993. The announcement included a joint specification that has three elements -- 1. An API, Application Programming Interface for applications developers. 2. A SPI, Service Provider Interface for service providers and network interface cards. 3. Windows DLL, software code to be incorporated into Microsoft Windows.

WINK
A momentary interruption in SF (single frequency) tone indicating the distant central office is ready to receive the digits you just dialed.

WINK RELEASE
On most modern central offices when the person or device at the other end hangs up, your local central office will send you a single frequency tone. That tone is called wink release. Such a tone can be used to alert a data device that the device at the other end has hung up. (Remember it can't tell by just listening -- like you and me.) When a data device hears a wink release, it usually takes it as a signal to hang up also.

XENIX
Microsoft trade name for a 16-bit microcomputer operating system derived from AT&T Bell Labs' UNIX.

YELLOW ALARM
A T-1 alarm signal sent back toward the source of a failed transmit circuit in a DS-1 2-way transmission path. A yellow sends 0's (zeros) in bit two of all time slots. See also T-1.

YELLOW PAGES
A directory of telephone numbers classified by type of business. It was printed on yellow paper throughout most of the twentieth century until it was obsoleted in the late 1990s by dial-up yellow page directories operated by voice processing systems and in the early 21st century by electronic directories delivered on disposable laser disks. As a concession to history,

the laser disks are now painted bright yellow. Actually, yellow pages remain one of the phone companies' most lucrative sources of revenues. Advertising rates are not cheap. There is now competition. There are many "Yellow Pages" directories, since AT&T never trademarked the term "Yellow Pages." Some "yellow page" directories are better value than others. And some are more legitimate than others. Some actually never get printed or, if they are printed, are not printed in great quantity and are not distributed as widely as their sales literature implies. Many businesses have been suckered into paying money for listings and advertisements in directories that never appeared. This fictitious directory scam also has happened with "telex" and "fax" directories. This "scam" is fraud by mail and is heavily stomped upon by the US Postal Service. As a result, many fake directories (especially the telex ones) are "published" abroad.

Z

ZERO CODE SUPPRESSION

The insertion of a "one" bit to prevent the transmission of eight or more consecutive "zero" bits. Used primarily with digital T1 and related telephone-company facilities which require a minimum "ones density" keep the individual subchannels of a multiplexed, high-speed facility active. Several different schemes are currently employed to accomplish this. Proposals for a standard are being evaluated by the CCITT. See also ZERO SUPPRESSION.

ZIP TONE

Short burst of dial tone to an ACD agent headset indicating a call is being connected to the agent console.